RAND ARROYO CENTER

Active Component Responsibility in Reserve Component Pre- and Postmobilization Training

Ellen M. Pint, Matthew W. Lewis, Thomas F. Lippiatt,
Philip Hall-Partyka, Jonathan P. Wong, Tony Puharic

T0308384

Prepared for the United States Army
Approved for public release; distribution unlimited

For more information on this publication, visit www.rand.org/t/rr738

Library of Congress Control Number: 2015938409

ISBN 978-0-8330-8783-6

Published by the RAND Corporation, Santa Monica, Calif.

© Copyright 2015 RAND Corporation

RAND® is a registered trademark.

Support RAND

Make a tax-deductible charitable contribution at
www.rand.org/giving/contribute

www.rand.org

Preface

This report describes the results of a project entitled Active Component Responsibility in Reserve Component Pre- and Postmobilization Training. This project examined the laws and policies governing active component (AC) support for reserve component (RC) training to meet predeployment training requirements and/or Army Force Generation training goals and identified changes in legislation and Army policy needed to provide the required level of support in the future.

This document provides an overview of the evolution of AC support for RC training from 1973 to the present. It examines the congressional intent behind Title XI of the National Defense Authorization Act for Fiscal Year 1993, which mandates the number of AC personnel to be assigned to support RC training, among other provisions designed to improve RC readiness and training. This document also describes the Army's recent experience preparing RC units to support operations in Iraq and Afghanistan and its future plans for RC training requirements and training support. From this analysis, we recommend changes to Title XI and other policies to support the Army's future RC training plans.

This research was sponsored by the Director of Training in the Office of the Deputy Chief of Staff, G-3/5/7, U.S. Army, and conducted within the RAND Arroyo Center's Manpower and Training Program. RAND Arroyo Center, part of the RAND Corporation, is a federally funded research and development center sponsored by the United States Army.

The Project Unique Identification Code (PUIC) for the project that produced this document is RAN136452.

If you have any questions or comments regarding this report, please contact the project leader, Ellen Pint, at (310) 393-0411, extension 7529, or pint@rand.org.

Contents

Figures and Tables

Figures

Tables

Summary

To meet predeployment and Army Force Generation training requirements, Army reserve component (RC) units need support from other
Army organizations, such as First Army, other active component (AC)
units, and RC training support brigades and higher headquarters. Legislation passed in the early 1990s, in response to readiness problems
in RC units mobilized to support Operation Desert Storm, mandates
the number of AC personnel assigned to support RC training and sets
other requirements for RC personnel and training. Since that time,
RC training policies and training support organizations have evolved
to meet the rotational demands for RC units to deploy to Iraq and
Afghanistan. As these operations come to a close and defense budgets
decline, the Army must determine what types of training support RC
units will need and how best to provide that support. To the extent that
these future arrangements differ from current law, the Army may need
to propose legislative changes to Congress.

This document describes the results of a research project examining AC support for pre- and postmobilization training of RC units. As
part of this project, we reviewed the historical context of AC support
for RC training and the congressional intent behind existing laws, analyzed predeployment training requirements and accomplishments for
RC units that deployed to Iraq and Afghanistan from 2003 to 2010,
and conducted interviews with training support providers and other
Army headquarters organizations to obtain information on the Army's
evolving plans for future RC training requirements and training support. From this research, we recommend changes to laws and policies
needed to support future RC training plans.

The Historical Context of Title XI

The main piece of legislation governing AC support for RC training was passed as Title XI of the National Defense Authorization Act for Fiscal Year 1993, also known as the Army National Guard Combat Readiness Reform Act of 1992. It was developed in response to the experience of mobilizing and training three Army National Guard (ARNG) roundout brigades to support Operation Desert Storm. Under the roundout concept, these brigades should have deployed with their parent AC divisions, which also included two AC brigades. However, the roundout brigades were not activated until late November 1990, four months after Desert Shield began, and required 90 to 135 days of postmobilization training. Combat operations had ended by the time the first brigade was validated for deployment.

Because these mobilizations were seen as a test of the roundout concept, the Department of the Army Inspector General observed and assessed the mobilization process, and its recommendations influenced subsequent reports by the U.S. General Accounting Office (GAO)[1] and the Congressional Research Service, as well as the congressional testimony of Army leaders. The major provisions of Title XI can be traced to specific problems related to the mobilization of the roundout brigades.

The problems the Department of the Army Inspector General (1991, pp. 1–3) identified included the following:

- Premobilization training lacked focus. Units did not meet expected levels of individual, crew, and platoon proficiency. Postmobilization training plans had to be adjusted to provide sufficient time to retrain and attain the prescribed standards.
- The ARNG brigades had serious personnel readiness problems, including low manning levels of critical combat arms and low-density support specialties; military occupational specialty qualification shortfalls, and lack of medical and/or dental preparedness.

---

[1]   Now known as the U.S. Government Accountability Office.

- Many officers and noncommissioned officers in key positions were ineffective in performing their duties. Leaders had not attended required professional development courses or lacked experience in their positions.
- Expectations of initial levels of training and readiness in the roundout brigades were too high because of inadequate measures and procedures for determining premobilization readiness.

The provisions of Title XI that were intended to address these problems included the following:

- assigning 5,000 AC advisers to RC units (later reduced to 3,500 in 2005) to increase the quantity and quality of full-time support
- associating each ARNG combat unit with an AC combat unit and giving the AC commander (at brigade or higher level) the responsibility to approve the ARNG unit's training program, review its readiness reports, assess its resource requirements, validate its compatibility with AC forces, and approve vacancy promotions of officers
- establishing a program to minimize postmobilization training time by focusing premobilization training on individual soldier qualifications and training, collective training at the crew or squad level, and maneuver training at the platoon level
- modifying the RC readiness rating system to provide a more accurate assessment of deployability and personnel and equipment shortfalls that require additional resources
- setting an objective of increasing the percentage of ARNG personnel with prior AC experience to 65 percent for officers and 50 percent for enlisted.

The Army's implementation of Title XI was integrated into Bold Shift, a pilot program to improve RC training that was established by the National Defense Authorization Act for Fiscal Years 1992 and 1993, which initially required 2,000 AC advisers to be assigned to RC units. Bold Shift refocused collective training for RC combat units at the crew, squad, and platoon levels. AC advisers and AC associate units provided the opposing force, observer-controllers, crew examiners, and other support personnel so that the RC unit could focus on training and increase the number of tasks trained. However, evaluations of Bold Shift by the RAND Arroyo Center (Sortor et al., 1994) and GAO (1995) found that, even with the additional resources dedicated to RC training and more-limited training goals, RC units could not complete individual, crew, and maneuver training because of the limited total RC training time available and the low attendance rates at annual training.

In 1996 and 1997, the Army reviewed its RC training support organizations and recommended the creation of a tricomponent organization with a single chain of command. These changes increased the role of the Continental U.S. Armies (First Army and Fifth Army) and reduced the role of AC associate units in RC training support. In addition to assigned Title XI AC personnel, the Continental U.S. Armies were given operational control over Army Reserve Training Support Brigades. Concurrently, Congress allowed the Army to count AC personnel assigned to units with the primary mission of providing training support to RC units as part of the total number of AC advisers required by Title XI. Under a tiered readiness concept, however, training support resources were focused on units that had a high priority for mobilization, while less support was provided to lower-priority units.

Evolution of Pre- and Postmobilization Training to Support Operations in Iraq and Afghanistan

As RC units started to be mobilized in 2001 to support operations in Iraq and Afghanistan, most of the Soldier Readiness Processing (SRP), individual soldier qualification and training, and collective training were conducted after mobilization. RC units that deployed initially had relatively short notice, but even as notification periods increased, high rates of personnel turnover in the year before deployment still made it difficult to schedule required training more than a few months before mobilization. However, in January 2007, the Secretary of Defense issued a memorandum limiting involuntary mobilizations to a maximum of one year. This policy change forced the Army to shift more training to the premobilization period to maximize the amount of time RC units could spend in theater. Since First Army had been focusing on postmobilization training support for RC units, the ARNG and U.S. Army Reserve (USAR) increased premobilization training support, establishing guard Premobilization Training Assistance Elements in each state and three reserve Regional Training Centers (one each in California, New Jersey, and Wisconsin).

We examined the number of pre- and postmobilization training days selected types of RC units needed to prepare for deployment, using a database developed for other studies at RAND. Comparing units mobilized in 2003–2007 with those mobilized in 2008–2010, we found that, in response to the one-year limit on mobilization time, the Army was able to reduce postmobilization training time by an average of 20 to 35 percent, depending on unit type. Although the total number of pre- and postmobilization training days needed declined for some unit types (including brigade combat teams preparing for counterinsurgency missions and support units that did not have to travel frequently off forward operating bases), they remained constant for other unit types. Premobilization training focused on SRP and individual soldier requirements. However, in most RC units, only 70 to 80 percent of soldiers were able to complete all required tasks, for such reasons as lack of access to the most up-to-date equipment (such as body armor, night-vision equipment, weapons, and vehicles), annual training attendance rates of 70 to 80 percent, and cross-leveling of new soldiers into deploying units in the last few months before mobilization.

We also examined a small sample of after-action reviews that RC units submitted to First Army approximately 90 days after arrival in theater. Some of the concerns these reports expressed were similar to those of AC units, such as availability of theater-specific equipment to train on and the relevance of predeployment training to current conditions in theater. Others were more specific to RC status, including the difficulty of achieving premobilization medical and dental readiness, lack of access to and training on AC automated systems and databases, and repetition of training events at the mobilization station that had already been completed at home station.

Thus, many of the same problems observed in the roundout brigades in 1990 and 1991 persisted in RC units preparing to deploy to Iraq and Afghanistan, although some were less severe. These problems included a limited number of premobilization training days; limited access to training ranges, maneuver areas, and some types of equipment; and personnel turnover and annual training attendance issues that limited the effectiveness of premobilization training.

Implications of Army Plans for Future RC Training

Based on our interviews with Army training support providers and other headquarters organizations, the Army plans to continue its Army Force Generation readiness process for RC forces on a 60-month cycle. Units that are needed to meet known rotational requirements or to execute the first rotation of an operation plan will be assigned to the Rotational Force Pool, while those that are not needed until the second or third rotation will be assigned to the Operational Sustainment Force Pool. In each five-year cycle, RC units in the Rotational Force Pool will spend up to one year in the reset phase and three years in the train/ready phase and will be available for known deployments or contingency operations for one year. Training plans suggest that brigade combat teams would focus primarily on collective training activities during the train/ready phase and achieve company-level live-fire and maneuver proficiency by the end of that phase.

However, past experience with Bold Shift and RC units preparing for more-recent deployments indicates that, even with additional training resources, RC units struggled to complete individual and crew- or squad-level training and achieve platoon-level proficiency. Furthermore, SRP and other individual training requirements not met prior to mobilization will have to be completed after mobilization, which would affect the types of postmobilization training support needed.

First Army estimates that it will need to support training events for 70,000 RC soldiers in the Rotational Force Pool each year. To fully support this throughput of soldiers, it needs about 3,000 trainer-mentors but expects to have fewer than 2,000 because some Title XI

AC personnel will be assigned to command and support positions in First Army and other Army commands. During recent operations, First Army relied on mobilized reserve personnel and temporary civilian hires, but funding for these positions is coming to an end. In peacetime, it may be able to increase usage of USAR training support personnel during their annual training periods to support training events for RC units.

Conclusions and Recommendations

Historical evidence and more-recent experience suggest that premobilization training should focus on individual soldier qualifications and training and collective training at the crew, squad, and platoon levels, particularly for combat units. Some company-level training may be feasible for enabler units and, as time permits, for combat units. The Army currently has a multicomponent RC training support structure that has worked relatively well in support of recent operations. It is important to maintain unified, multicomponent training support organizations, to be consistent with Total Force Policy, to ensure that training standards do not diverge across components in the future, and to conserve resources in a time of declining budgets. However, First Army may need to make greater use of USAR training support personnel during their annual training periods to support premobilization collective training events, and the ARNG could increase its involvement by filling its authorized positions in First Army.

Finally, First Army's after-action review process could be expanded and improved to provide better feedback and inform process improvement during peacetime, as well as during any future large-scale mobilizations. Its feedback mechanisms could be strengthened by using standardized questions to make data comparable across units and over time, spreading out the feedback process over time to reduce the postdeployment reporting burden, and making the results of the feedback process more easily accessible to First Army planners and to units going through the mobilization process.

While some provisions of Title XI remain relevant, others no longer reflect the current operating environment. Whether the remainder should be retained depends on the expense or difficulty of compliance and how the Army decides to structure its future training support organizations. It is still important for the AC to be involved in RC training, but since the Army force structure is changing, it is not clear exactly how many AC personnel should be assigned to this role. A more-flexible approach might be to specify the proportion of AC personnel assigned as trainer-mentors or elsewhere to support RC training. The Title XI requirement for RC units to be associated with similar AC units has become outdated; AC-led multicomponent units, such as First Army, now fulfill the roles and responsibilities of AC associate units. Goals for the percentage of officers and enlisted with AC experience are less relevant, given the large fraction of RC personnel with recent deployment experience. However, many provisions could be retained, including requirements to focus premobilization training at the individual, crew, squad, and platoon levels; improve the accuracy of readiness ratings; and increase the compatibility of AC and RC equipment and automated systems.

Acknowledgments

This project benefited from the support and guidance of its sponsor, BG Todd McCaffrey, former Director of Training in the Office of the Deputy Chief of Staff, G-3/5/7. Special thanks are given to our action officers, LTC Chester Guyer and MAJ Shawn Dillon in the Office of the Deputy Chief of Staff, G-3/5/7, DAMO-TRC, who provided thoughtful stewardship of our efforts and access to many important resources and contacts that greatly added to the quality and timeliness of our research products.

The research was also strengthened by site visits, interviews, and teleconferences with Army unit personnel and civilians in the Office of the Deputy Chief of Staff (G-3/5/7), Office of the Assistant Secretary of the Army for Manpower and Reserve Affairs, U.S. Army Forces Command, the National Guard Bureau, U.S. Army Reserve Command, First Army, and the 196th Infantry Brigade, whom we thank without specific note of their positions to maintain their anonymity. Their candid input was a key part of understanding the details of RC training and mobilization processes, resource demands, unmet demands, and constraints on training and mobilization activities. We owe them a large debt of gratitude.

At RAND, we benefited from feedback from many colleagues, but our work was particularly strengthened by the critiques of Michael Hansen, director of the Arroyo Center Manpower and Training program, throughout the project. We would also like to thank the reviewers of this document, James Boling of RAND and Stanley Horowitz of the Institute for Defense Analyses, for their helpful comments. In addition, we thank Martha Friese, Joan Myers, and Donna White for their able assistance in all phases of this research.

CHAPTER ONE

Introduction

To prepare to support combatant commander requirements, Army National Guard (ARNG) and U.S. Army Reserve (USAR) units typically need support from various external sources—such as First U.S. Army, other active component (AC) units, and reserve component (RC) training units and higher headquarters. Legislation dating back to the early 1990s established goals and requirements for RC personnel and training, AC support to RC units, and reporting requirements to Congress. However, in recent years Army structure, missions, and force generation processes have evolved in ways that affected how RC unit readiness is supported. The experience of nearly a decade of RC mobilizations has changed training practices and produced numerous lessons for preparing RC forces efficiently. As operations in Afghanistan come to an end, budgets decline, and the new Army Total Force Policy is implemented, the Army must determine the types of training support RC units need across the Army Force Generation (ARFORGEN) cycle and how best to provide that support. The current Army training strategy establishes unit readiness aim points and unit proficiency levels across the ARFORGEN cycle based on force pool assignment and planned operational missions. To the extent that future training needs differ from current law and policy, changes may be needed in legislation and Army policy, regulations, training practices, and culture.

Therefore, the Director of Training in the Office of the Deputy Chief of Staff, G-3/5/7, asked RAND Arroyo Center to examine the historical evolution of pre- and postmobilization training and training support requirements and the Army's planned future training strategy for RC units to identify the types of support these units need to meet pre- and postmobilization training requirements and/or to achieve ARFORGEN training aim point goals. The Director of Training also asked us to recommend changes in law, policy, and regulations to provide the required level of support. This research initially involved four tasks:

1. **Identify key types of RC units needed to support combatant commander requirements.** In coordination with the study sponsor, we selected a cross section of combat, combat support (CS), and combat service support (CSS) unit types that have been used in recent operations or would be needed to support defense planning scenarios.

2. **Document the historical evolution of pre- and postmobilization training and training support requirements for key RC unit types.** For selected unit types, we examined how training support for RC units evolved to support mobilization requirements and evaluated data on the number of pre- and postmobilization training days needed for each RC unit type and mission. We also examined changes in the types of support First Army and other AC and RC trainers provide.

3. **Examine implications of future employment requirements for RC training and training support requirements.** The research team gathered information on operation plans (OPLANs), theater support plans, the assignment of RC units to force pools, and ARFORGEN training aim point goals that affect the capabilities and capacities necessary to meet RC training support needs during interviews with representatives from Army G-3, U.S. Army Forces Command (FORSCOM), First Army, U.S. Army Reserve Command (USARC), and the National Guard Bureau (NGB).

4. **Compare current legal requirements with current and expected future RC training support needs and recommend potential legislative changes.** The research team reviewed existing legislation on AC support of RC training and identified sections of the law that have become outdated as force genera-

tion processes and RC training support have evolved to meet rotational demands for forces. We recommend that the Army propose changes to these provisions to reflect expected future needs for AC training support of RC units.

However, in consultation with the sponsor, the emphasis of the study shifted to support the Army's ongoing decision processes for RC training strategy and training support requirements. In March 2013, the Secretary of the Army established the Total Army Training Validation Integrated Planning Team, cochaired by the Deputy Assistant Secretary of the Army (Training, Readiness and Mobilization) and Department of the Army Military Operations–Training, to establish a Total Army framework for collective training oversight under which commanders certify their training proficiency and readiness; their higher commanders confirm the assessments; and an independent organization validates these assessments for the Secretary of the Army (McHugh, 2013). The sponsor requested that we provide information to the team regarding the congressional intent underlying the legislation mandating AC support of RC training, i.e., Title XI of the National Defense Authorization Act (NDAA) for Fiscal Year (FY) 1993 (Public Law [PL] 102-484), also known as the Army National Guard Combat Readiness Reform Act of 1992.

Some additional questions also arose as a result of our interviews with training providers and other headquarters organizations, including FORSCOM, USARC, Army G-3, NGB, and First Army. In particular, given the shift in First Army's role from supporting premobilization training to supporting postmobilization training for operations in Iraq and Afghanistan and greater ARNG and USAR involvement in supporting premobilization training, there was some debate about the relative roles of the AC, ARNG, and USAR in supporting RC training. Finally, our analysis of recent experience with pre- and postmobilization training accomplishments raised questions about the timing of individual and collective training in the ARFORGEN cycle. For example, to the extent that Army and theater-specific individual training requirements and Soldier Readiness Processing (SRP) cannot be completed during premobilization training, these requirements are shifted to the postmobilization period, with implications for the type and capacity of postmobilization training support that will be needed.

Thus, the focus of the study shifted away from a quantitative assessment of the personnel and resources needed to provide training support under various conditions to a more qualitative evaluation of the historical roles of the AC, ARNG, and USAR in preparing RC units for deployment and an assessment of how these roles might evolve in the future.

Research Methodology

Three main research efforts were involved in this study. First, we conducted a review of public laws and U.S. Code (USC) associated with Title XI and subsequent revisions. We also obtained congressional reports and transcripts of hearings, as well as other reports that influenced the development of the law, such as the Army Inspector General's assessment of the National Guard roundout brigades that were mobilized for Operation Desert Shield–Operation Desert Storm (ODS) (Department of the Army Inspector General, 1991) and a Congressional Research Service report on the roundout brigades (Goldich, 1991). In addition, we reviewed the literature describing Army and national policies regarding use of RC forces and AC support of RC training from the 1970s to the present.

Second, we used databases that the RAND Arroyo Center and National Defense Research Institute had developed, supplemented by other information on predeployment training requirements and pre- and postmobilization training accomplishments, to examine the Army's more-recent experience preparing RC units to deploy to Iraq and Afghanistan from 2003 to 2010. This analysis included an assessment of the number of premobilization and postmobilization training days selected types of RC combat, CS, and CSS units needed to prepare for deployment. It also examined how pre- and postmobilization training evolved after the Secretary of Defense issued a 2007 memorandum that limited each RC mobilization to one year.

Third, we conducted interviews with representatives from training providers, including First Army and the 196th Training Support Brigade (TSB), and other Army headquarters organizations, including Army G-3, FORSCOM, USARC, and NGB. These interviews focused on such topics as

- evolution of training requirements and training support during recent operations
- planned future training requirements as operations in Afghanistan come to an end, including assignment of RC units to ARFORGEN force pools, pre- and postmobilization training requirements and ARFORGEN training aim points, and OPLAN surge requirements
- types of training support that will be needed in the future, including trainer-mentors and facilities.

To complement these interviews, we also examined First Army after-action reviews (AARs) and Center for Army Lessons Learned (CALL) documents to obtain comments from RC units on the effectiveness of predeployment training.

Outline of This Report

In Chapter Two, we briefly review the history of policies on utilization of RC units and AC support for RC training. We then provide a more detailed description of the major provisions of Title XI, the specific congressional concerns and issues underlying these provisions, and the subsequent evolution of AC support for RC training in the 1990s. Chapter Three describes the further evolution of pre- and postmobilization training to support Operation Enduring Freedom (OEF) and Operation Iraqi Freedom (OIF), including our analysis of pre- and postmobilization training days by unit type. Chapter Four discusses the implications of Army plans for post-OEF RC training requirements for AC support of RC training, and Chapter Five summarizes the conclusions and recommendations arising from the study. The appendix provides excerpts from relevant legislation.

CHAPTER TWO
The Historical Context of Title XI

This chapter first provides an overview of events that influenced AC support of RC training from the 1970s to the time of our research. The first section establishes the broad historical context for Title XI and subsequent policy changes. The second section describes the major provisions of Title XI, as well as the specific concerns that Congress was trying to address, which related to the mobilization of three National Guard roundout brigades to support ODS. The third section discusses the Army's implementation of Title XI and changes in AC support of RC training in the remainder of the 1990s.

Overview: 1973 to 2013

The time lines in Figures 2.1 and 2.2 summarize the key events that influenced the role of the RCs and AC support for RC training. Following the end of the Vietnam War and the advent of the All-Volunteer Force, the Department of Defense (DoD) implemented the Total Force Policy, which established that all military assets, including the ARNG and USAR, should be treated as a single integrated force.[1] Secretary of Defense James Schlesinger stated that the "Total Force Policy integrates the Active, Guard, and Reserve Force into a homogeneous whole" (Broomall, 1992). As the AC experienced reductions

---

[1] This policy was in contrast to the Vietnam War, which was conducted primarily with AC forces and draftees. With a very few exceptions, President Lyndon Johnson refused to mobilize the RCs (MacCarley, 2012, pp. 38–39).

Figure 2.1
Time Line of AC Support to RC Training (1973 to 1993)

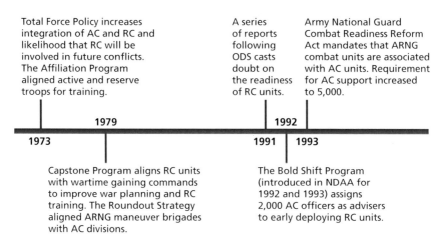

Total Force Policy increases integration of AC and RC and likelihood that RC will be involved in future conflicts. The Affiliation Program aligned active and reserve troops for training.

A series of reports following ODS casts doubt on the readiness of RC units.

Army National Guard Combat Readiness Reform Act mandates that ARNG combat units are associated with AC units. Requirement for AC support increased to 5,000.

1979   1992

1973   1991 1993

Capstone Program aligns RC units with wartime gaining commands to improve war planning and RC training. The Roundout Strategy aligned ARNG maneuver brigades with AC divisions.

The Bold Shift Program (introduced in NDAA for 1992 and 1993) assigns 2,000 AC officers as advisers to early deploying RC units.

RAND RR738-2.1

during the 1970s, war plans increasingly relied on the capabilities of RC forces. The Affiliation Program was also approved in 1973 as a way to improve RC readiness by fostering stronger relationships between AC and RC units. At first, the program only applied to combat arms units but was later expanded to early deploying CS and CSS units in 1976 (Arnold, 2003).

Under the leadership of Army Chief of Staff GEN Creighton Abrams, the Army adopted the Roundout Strategy, which designated ARNG maneuver brigades as one of the three combat brigades in several AC divisions. The Roundout Strategy was intended both to increase the total number of Army divisions without increasing AC Army end strength and to improve the readiness and visibility of the Army reserve components by assigning them higher-profile missions. The Army also gave the roundout brigades higher priority to receive modernized weapons and equipment (Goldich, 1991, and Buchalter and Elan, 2007).

The CAPSTONE program was launched in 1979 to align RC units with the AC or other RC units with which they would likely be employed in wartime. Each reserve unit was designated a wartime chain of command, probable wartime mission, and probable area of employ-

Figure 2.2
Time Line of AC Support to RC Training (1994 to 2013)

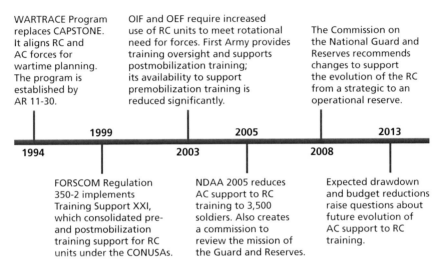

ment based on existing combatant command OPLANs. FORSCOM was directed to implement CAPSTONE and tasked the three Continental U.S. Armies (CONUSAs) with managing the reserve units in their assigned geographical regions. RC units were intended to tailor their training plans to their expected wartime missions and participate in joint training exercises with other aligned units. Three years later, a 1982 U.S. General Accounting Office (GAO)[2] report found that many RC units had neither been contacted by their wartime gaining commands nor received training and planning guidance. GAO also found that there was no formal reporting system to monitor CAPSTONE implementation and that FORSCOM and the CONUSAs were unable to determine whether RC units were receiving the required training and planning guidance (GAO, 1982).

The first major test of the Total Force Policy was ODS in 1990 and 1991. By most measures, the mobilization of RC support units (such as transportation, medical, engineering, and military police) for

---

[2] Since renamed the U.S. Government Accountability Office.

ODS was considered a success, although the Army recognized that there were areas for improvement in personnel readiness, including military occupational specialty (MOS) qualification and medical and dental readiness (Sortor et al., 1994, p. 2). Across all the services, a total of 228,000 reservists were mobilized, of whom 140,000 were USAR or ARNG members. Army RC soldiers and their units performed a crucial role in CS and CSS because most of this capability resided in the RC. However, the 1st Cavalry and 24th Infantry Divisions were deployed without their ARNG roundout brigades. Under the roundout concept, one of the three brigades of several AC Army divisions was an ARNG brigade. In 1990, seven of the Army's active divisions had roundout brigades, and another three had roundout battalions (Goldich, 1991, pp. 5–7).

After Iraq invaded Kuwait in August 1990, the commander of U.S. Central Command requested two full-strength heavy divisions. At the time, these divisions faced the possibility of immediate combat on arrival in the theater of operation to defend Saudi Arabia from an Iraqi invasion, so it was thought that there would not be enough time for postmobilization training of their roundout brigades. As a result, active brigades were substituted for the roundout brigades in these divisions. Moreover, the reserve call-up authority the President invoked in August 1990 allowed reservists to be kept on active duty for a maximum of 180 days, which would not allow enough time for postmobilization training. However, as Iraq continued to build up its forces in Kuwait, the requirements for U.S. forces increased, as did institutional and political pressures to activate the roundout brigades. Several influential members of Congress supported activation, including Les Aspin, then the chairman of the House Armed Services Committee. The NDAA for 1991 included a provision expressing the sense of Congress that the President should activate at least one roundout brigade (Goldich, 1991, pp. 9–13).

Two mechanized infantry brigades (the 48th Infantry Brigade of Georgia and the 256th Infantry Brigade of Louisiana) were activated on November 30, 1990, and an armored brigade (the 155th Armored Brigade of Mississippi) on December 7, 1990. Initially, the ARNG brigade commanders estimated that their units needed 40 days of postmobilization training to be combat ready, based on their readiness reports. Second Army and III Corps raised this estimate to over 90 days, based on their assessment of the brigades' proficiency. According to the GAO (1991, p. 3),

> many soldiers were not completely trained to do their jobs; many noncommissioned officers (NCOs) were not adequately trained in leadership skills; and Guard members had difficulty adjusting to the active Army's administrative systems for supply and personnel management. … Also, when activated, many soldiers had serious medical or dental conditions that would have delayed or prevented their deployment.

Postmobilization training plans had underestimated the amount of training that would be needed because peacetime evaluations had overstated the brigades' readiness.

The active Army validated the 48th Brigade as being ready for deployment on February 28, 1991, the date of the cease-fire with Iraq, after approximately 90 days of postmobilization training. The 155th Brigade was scheduled to complete validation on March 22, 1991 (105 days after activation) and the 256th Brigade on April 13, 1991 (135 days after activation). These longer training periods were due to the limited availability of training facilities and personnel and the need to train the 256th Brigade on its newly issued Bradley Fighting Vehicles rather than to differences in the quality of personnel or the premobilization readiness of the brigades (Goldich, 1991, pp. 11, 13–14).

Because the activation of the three ARNG brigades was seen as a test of the roundout concept, the Chief of Staff of the Army tasked the Army Inspector General to observe and assess the efficiency of the mobilization and training of the three brigades. Dedicated teams, headed by former AC combat brigade commanders, followed and observed the actions of brigades and their AC trainers from alert through demobilization. The congressional testimony we reviewed reflected the conclusions and recommendations of the Department of the Army Inspector General's (DAIG's) report (1991), which had a strong influence on the provisions of Title XI. The GAO and Congressional Research Service also produced reports on the mobilization and training of the roundout brigades. The next section discusses the findings of these reports and hearings in more detail and describes how they led to the provisions of Title XI.

To improve the readiness of the roundout brigades, the Army developed a set of initiatives called Bold Shift, which the Chief of Staff approved in the fall of 1991. Seven ARNG roundout brigades participated in the initial introduction of Bold Shift in 1992, along with an additional 82 nondivisional support elements from the ARNG and USAR. Bold Shift refocused RC training on crew-, squad-, and platoon-level proficiency for combat units and company-level proficiency for CS and CSS units (Sortor et al., 1994). Concurrently, the NDAA for FYs 1992 and 1993 (PL 102-190, passed in December 1991) required the Army to establish a pilot program to assign at least 1,300 AC officers as advisers to RC combat units and 700 officers as advisers to RC CS and CSS units that had a high priority for deployment. These advisers were to be assigned to full-time duty in connection with organizing, administering, recruiting, instructing, and training these units.

The following year, Congress passed the Army National Guard Combat Readiness Reform Act of 1992, which was designated as Title XI of the NDAA for FY 1993. Title XI included nearly 20 provisions designed to improve the readiness, training, and deployability of ARNG combat units. Among these provisions was a requirement that each ARNG combat unit be associated with an AC combat unit whose commander (at brigade level or higher) would approve the ARNG unit's training program; review its readiness report; assess its manpower, equipment, and training resource requirements; validate its compatibility with AC forces; and approve vacancy promotions of officers.[3] It also required assignment of an additional 3,000 AC warrant officers and enlisted soldiers as advisers to RC units, raising the total from 2,000 to 5,000 (PL 102-484, 1992).

---

[3] Title XI did not originally specify the echelon at which ARNG and AC combat units would be associated, but the NDAA for FY 1996 modified this section to refer to "each ground combat maneuver brigade for the Army National Guard that (as determined by the Secretary) is essential for the execution of the National Military Strategy."

As Figure 2.2. shows, AC support for RC training continued to evolve in the 1990s. The CAPSTONE program was replaced by WARTRACE in 1994. Under WARTRACE, combatant commanders developed OPLANs for specific contingencies in their areas of responsibility. FORSCOM would then identify specific units, including RC units, for inclusion in these OPLANs. These units would be assigned a wartime chain of command, which for RC units was usually different from the peacetime chain of command. The wartime chain of command would provide guidance for the unit's premobilization training program (Army Regulation [AR] 11-30, 1995, and Chapman, 2008). FORSCOM Regulation 350-2, "Reserve Component Training," was revised to reflect changes in premobilization training requirements, expected training proficiency levels, and the training support that RC units would receive. Importantly, the regulation gave the FORSCOM commander responsibility to develop training criteria and to distribute AC resources to assist in RC training, while the director of the ARNG and the commander of USARC retained responsibility for establishing training policies to meet premobilization training requirements (FORSCOM, 1999).

FORSCOM conducted a formal evaluation of RC training support in 1996 and began to implement its recommendations, known as Training Support XXI, in 1997. Training Support XXI consolidated pre- and postmobilization training support under the remaining CONUSAs, First Army and Fifth Army. Training support organizations that were previously controlled by the CONUSAs and USARC were reorganized into TSBs under the operational control of the CONUSAs. The role of AC units in supporting the training of their associated RC units was also transferred to the CONUSAs. Training Support XXI was fully implemented in October 1999 (Arnold, 2003).

The ARNG and USAR have played an important role in meeting the rotational demand for forces to support OEF and OIF/ Operation New Dawn (OND). The total number of mobilized RC service members, across all services, peaked at over 200,000 in May 2003 and remained at around 100,000 through 2007, with about two-thirds coming from the ARNG and USAR (Defense Science Board Task Force, 2007). In 2004, more than one-third of all U.S. military forces (from all services) in Iraq were from the RC (Commission on the National Guard and Reserves, 2008). Reacting to this demand, First Army shifted its resources to supporting postmobilization training for RC units preparing to deploy. However, as a result of this shift, fewer resources were available for premobilization training.

To create a more modular and flexible structure that would better support the rotational demands for forces in OEF and OIF, the Army also transformed from a division-based to a brigade-based structure. In the NDAA for FY 2005, Congress allowed the Army to reduce the number of AC advisers from 5,000 to 3,500 and to reallocate officer and NCO positions to modular brigades, while expanding the number of reservists on active duty to support postmobilization training of RC units. The NDAA for FY 2005 also created the Commission on the National Guard and Reserves to assess the roles and missions of the reserve components and to recommend changes to improve readiness and national security (PL 108-375, 2004).

The commission released its report in 2008. It found that there was no reasonable alternative to DoD's continued reliance on the RCs as part of its operational force and made 95 recommendations to improve RC personnel management, equipping, training, and other policies. Regarding training, the commission recommended that the services reassess the number of training days that RC units require prior to mobilization, because 39 days per year may not be adequate to meet the standards established by their force generation models. It also recommended that Army RC units be certified ready at the company level prior to mobilization and that Army organizations responsible for certification be engaged before mobilization to avoid repeated checks at postmobilization training sites (Commission on the National Guard and Reserve, 2008).

More recently, the end of OIF/OND, the drawdown in Afghanistan, and competing demands for federal funds have led to budget reductions across DoD, including the Army. Because First Army's postmobilization training support workload has declined, the Army must determine how to provide premobilization and postmobilization training support for RC units in the future. To the extent that planned future training support arrangements differ from existing law, including Title XI, the Army will need to request changes from Congress.

Major Provisions of Title XI

Table 2.1 summarizes the major provisions of Title XI of the NDAA for FY 1993, also known as the Army National Guard Combat Readiness Reform Act of 1992, and the related issues and concerns that prompted Congress to enact Title XI. A full list of the provisions of Title XI and subsequent revisions can be found in the appendix. As noted in the previous section, Title XI was largely a response to the experience of mobilizing and training three ARNG roundout brigades to support ODS.

According to the House Armed Services Committee's report on H.R. 5006, which became the foundation for the NDAA for FY 1993 (U.S. House of Representatives, 1992d, pp. 19–21), Congress intended Title XI to focus on six key areas:

- **Increasing experience and leadership in the ARNG.** Title XI would require that, by 1997, 65 percent of the new officer intake and 50 percent of the new enlisted personnel intake have at least two years' prior service. Commanders of active duty units associated with RC units would be required to review promotions of officers to captain and above. NCOs would be required to complete military education requirements prior to promotion to a higher grade.
- **Focusing and improving training.** Larger combat units are called on to display complex collective skills that cannot be maintained with 39 training days a year. Title XI would focus training on individual and small-unit skills, leaving larger-unit training for the period after mobilization.
- **Strengthening personnel standards.** Title XI would establish stricter medical, dental, and physical screening and create a special, nondeployable category for soldiers who do not meet physical or fitness standards.

Table 2.1
Major Provisions of Title XI and Underlying Issues

Title XI Provisions	Concerns in 1991–1992
§1132: 2,000 AC officers and 3,000 warrant officers and enlisted personnel to be assigned as advisers to RC units	More than 3,600 AC personnel were used to support postmobilization training of roundout brigades
1996: Amended to include personnel assigned to an AC unit providing dedicated training support to RC units 2005: Number of AC advisers reduced from 5,000 to 3,500	Need to increase quantity and quality of full-time support personnel, augment existing AGRs and military technicians (modeled on Marine Corps inspector-instructor program)
§1131: Each ARNG combat unit must be associated with an AC combat unit AC commander approves training program, reviews readiness reports, validates compatibility with AC forces, and approves vacancy promotions of officers	AC division commanders should have greater oversight of pre- and postmobilization training of ARNG roundout brigades Several company commanders were replaced due to lack of experience and/or leadership skills or ability
§1119: Establish a program to minimize postmobilization training time for ARNG combat units Premobilization training should emphasize: • Individual soldier qualification and training • Collective training and qualification at the crew, section, team, and squad levels • Maneuver training at the platoon levels and §1135: Identify priorities for mobilization of RC units and specify required number of postmobilization training days	RC units have limited time (39 days/ year) and resources for premobilization training Combat skills must be learned through repetitive field training in large maneuver areas that are not normally available for weekend training Existing postmobilization training plans for the roundout brigades were mostly scrapped after the brigades were activated because readiness deficiencies were greater than anticipated Postmobilization training time could be reduced by better integration of premobilization and postmobilization training plans and not devoting scarce premobilization training time to things best done after mobilization
§1121: Modify RC readiness rating system to provide a more accurate assessment of deployability and personnel and equipment shortfalls that require additional resources	Need for realistic expectations about roundout units' capabilities (e.g., not requiring them to deploy immediately for a rapid-response contingency) No single indicator that truly represents the readiness of a unit Specific concerns about MOS qualification of low-density MOSs and equipment shortages

Table 2.1—Continued

Title XI Provisions	Concerns in 1991–1992
§1120: Expand use of simulations, simulators, and advanced training devices and technologies	Need to increase use of simulation to make up for lack of training time and availability of ranges and large maneuver areas
	Provide battalion- and brigade-level commanders and staff training
	Increase training opportunities at weekend training locations
§1111: Increase percentage of ARNG personnel with prior AC experience to 65% for officers and 50% for enlisted[a]	Shortages of key personnel in roundout brigades
	Post–Cold War drawdown creates opportunities to increase proportion of RC soldiers with AC experience
§1114: Military education requirements for NCOs must be met for promotion to a higher grade	Large numbers of officers and NCOs had not attended required individual training courses
§1115: Establish a personnel accounting category for ARNG members who are not available for deployment	Up to 50 percent of soldiers failed initial physical tests and/or dental readiness standards
§1117: Annual medical and dental screening for each ARNG member (later repealed)	Demand for cross-leveling to replace nondeployable soldiers
§1134: Report on compatibility of AC and RC equipment and effects on combat effectiveness	Lack of organizational maintenance skills due to limited training time, availability of equipment at armories
	One roundout brigade had only recently received M-1 tanks and Bradleys and had not completed new equipment training
§1133: Ensure that personnel, supply, maintenance management, and finance systems are compatible across all components	RC automated systems (particularly personnel and supply management) were not compatible with AC systems
	Significant effort was needed to bring the Guard's policies and procedures into line with the AC's

[a] This provision was modified by the NDAA for FY 1996, §514, to a numerical requirement for 150 new officers and 1,000 new enlisted personnel with AC experience to join the ARNG each year, under specific programs. (See appendix for details.)

- **Removing impediments to effectiveness.** Title XI would require the Army to provide compatible automated systems for personnel, maintenance, supply, and finance for all Army components.
- **Creating new report cards.** Title XI would require the Army to modify its readiness rating system to assess unit deployability more accurately. Every ARNG combat unit would be associated with an active unit that would assess its training, readiness, and resource requirements.
- **Reforming the active Army.** For the ARNG to be effective in regional contingencies, the active Army must accept responsibility for the ARNG's readiness. Title XI would direct the Army to integrate the ARNG in its planning for regional contingencies and to allocate resources accordingly.

Although it is not specifically mentioned as one of the six key areas, section 1132 has received the most attention in recent years. It originally required the Army to assign 2,000 AC officers and 3,000 warrant officers and enlisted personnel as advisers to RC units, although congressional testimony and other sources do not provide a clear basis for these numbers. Goldich, 1991, p. 22, indicates that 3,600 soldiers from two AC mechanized infantry divisions, and more at the National Training Center and other posts, supported the roundout brigades' postmobilization training. The Army Inspector General's report says that about 2,800 AC personnel supported one of the roundout brigades and that a total of about 5,500 trainers were involved (DAIG, 1991, pp. 2-4 and 3-1), while GAO reported that nearly 9,000 AC personnel were assigned to train soldiers in the roundout brigades (GAO, 1991, p. 27). However, supporting postmobilization training involves different tasks from advising RC units on an ongoing basis.

The number of full-time support (FTS) personnel assigned to Army RC units (including active guard and reserve [AGR], AC personnel, and military technicians) was also lower than in the other services, as shown in Table 2.2.[4] Congressional testimony indicates that, at the

---

[4] FTS personnel perform a wide range of tasks that include organizing and supporting the training events, maintaining equipment, conducting administrative and record-keeping

Table 2.2
Percentage of Full-Time Support Personnel in
RC Units in 1991

Service	National Guard	Reserve
Army	12	8
Air Force	26	21
Marine Corps		15

time, about 8 percent of USAR personnel and 12 percent of ARNG personnel were providing FTS, while the Air Force Reserve and Air National Guard had 21 percent and 26 percent full-time personnel, respectively (U.S. House of Representatives, 1991a, p. 220). FTS for the Marine Corps Reserve was 15 percent of end strength, of which about 70 percent were AC personnel (U.S. House of Representatives, 1993a, p. 139). The Army Inspector General's report also cited inadequate quantity and quality of FTS personnel as a problem and noted that the "distribution of FTS personnel is often more at higher headquarters, where promotion opportunities are greater, than at battalion or company levels" (DAIG, 1991, pp. 2, 4–7).

Congressional testimony suggests that the original intent was to assign the AC advisers to RC units. In 1991, Representative Ike Skelton noted that the Marine Corps Reserve mobilized nine battalions that were able to get ready for deployment more quickly and participate in ODS. He attributed this performance to the sizable number of AC officers and experienced noncommissioned officers the Marine Corps assigned as instructor trainers in RC units (U.S. House of Representatives, 1991a, pp. 203–204).[5] COL James Davis, commander of the 48th Infantry Brigade, Georgia, agreed: "We need people who are going to get into slots and help us learn to fight, master gunners, maintenance

---

activities, and serving as recruiters and retention counselors. They allow drilling reservists to focus more of their time on training and readiness. (See Brauner and Gotz, 1991.)

[5] However, as Goldich, 1991, p. 42, notes, the degree of complexity involved in reaching and maintaining unit readiness is lower for battalions than for brigades.

people, people who can address our shortcomings" (U.S. House of Representatives, 1991a, p. 207).

In 1993 hearings focusing on implementation of Title XI, Representative Ike Skelton stated,

> One difference between the Army National Guard combat units and Marine Corps Reserve units is Active Duty oversight and control of the Marine Corps Reserve. ... [I]t seems clear that the Army could benefit from a program such as the Marine Corps' inspector/instructor program. ... Congress believed that increased Active Duty support was vital to Reserve component readiness. ... Congress intended that the personnel assigned be of high quality, and that their careers be protected while assigned to this important function." (U.S. House of Representatives, 1993a, p. 130)[6]

LtGen. Matthew Cooper, U.S. Marine Corps, Deputy Chief of Staff for Manpower and Reserve Affairs, described the program as follows: "Inspector-Instructors are Active Component Marines who supervise, instruct, and assist Selected Marine Corps Reserve ground units in attaining and maintaining a continuous state of readiness for mobilization" and "inspect and render technical advice to units in functions including administration, logistical support, and public affairs." They are "designed to ensure that the Reserve employed up-to-date Active Component training standards" (U.S. House of Representatives, 1993a, pp. 141–142).

However, the emphasis on assigning AC advisers to RC units faded after a few years. The NDAA for FY 1994 required the Army to establish one or more AC units with the primary mission of providing training support to RC units (PL 103-160, 1993, section 515), and the NDAA for FY 1996 allowed the Army to count AC personnel assigned to these units as part of the total number of AC advisers required by Title XI (PL 104-106, 1996). As noted in the previous section, the

---

[6] The NDAA for FY 1994 required the Army to submit an annual report on the implementation of Title XI as part of the Army Posture Statement, including a comparison of the promotion rates of officers assigned as AC advisers with those of all other Army officers (PL 103-160, 1993, section 517).

NDAA for FY 2005 reduced the required number of AC advisers from 5,000 to 3,500 (PL 108-375, 2004).

Another major provision of Title XI requires each ARNG combat unit to be associated with an AC combat unit. The commander of the AC unit, who must be at brigade level or higher, is responsible for approving the training program; reviewing the readiness report; assessing the manpower, equipment, and training resource requirements; and validating the compatibility of the ARNG unit with active duty forces. A related provision requires the commander of the associated AC unit to approve all officer unit vacancy promotions above the level of first lieutenant. The Army Inspector General had found that relationships between the roundout brigades and their parent AC divisions were not consistently strong. The effectiveness of the association between RC roundout units and AC divisions was based on the personal relationship between the RC brigade and AC division commanders and their commitment to the roundout concept. AC sponsor units did not always integrate roundout units into division training events and planning. Only one division commander consistently commented on the readiness of his roundout brigade in his unit status report (USR) (DAIG, 1991, pp. 2, 3, 4-4).

To address these problems, the Army Inspector General recommended that the AC division commander should approve and support the roundout unit's postmobilization training plan and rate the ARNG roundout brigade commander (DAIG, 1991, pp. 3, 4-12). In 1991 congressional testimony, the FORSCOM commander, GEN Edwin Burba, stated:

> During both pre- and postmobilization, it is highly desirable that training be conducted with and under the guidance of the AC division commander. This will ensure that standards are consistently validated throughout the roundout unit's training program. (U.S. House of Representatives, 1991a, pp. 175–176)

The Army Inspector General report also recommended that the Army improve the selection and training of roundout brigade leaders by centralizing the selection of brigade and battalion commanders and integrating the AC division commander into the process. In addition, many junior officers were not deployable because they had not completed the Officer Basic Course (DAIG, 1991, pp. 4-9, 4-10). Others were replaced because they lacked experience or leadership skills. BG Gary Whipple, commander of the 256th Infantry Brigade, Louisiana National Guard, testified:

> We had eight [company commanders] changed. ... Some of these young men had been in these command positions only a short period of time, had not gone through the modernization process with the Bradleys in particular ... . [W]e realized that this was probably above their experience level and their training level, so we replaced them. (U.S. House of Representatives, 1991a)

Some had more serious leadership deficiencies. LTG John Conaway, Chief of the NGB, stated: "[S]ome of them can look very good in drill status for that one weekend a month or three days a month, but not quite be what you want when you have them a longer period of time" (U.S. House of Representatives, 1991a, pp. 226–227).

Section 1119 of Title XI requires the Army to establish a program to minimize the postmobilization training time required for ARNG combat units. It requires unit premobilization training to emphasize individual soldier qualification and training; collective training and qualification at the crew, section, team, and squad levels; and maneuver training at the platoon level. Combat training for command and staff leadership is required to include multiechelon training to develop battalion-, brigade-, and division-level staff skills. The Army Inspector General report found that one of the reasons that postmobilization training times were so long for the roundout brigades was that units tried to do higher levels of collective training before becoming proficient in the lower levels. The report stated that these training shortfalls are correctable if premobilization and postmobilization training strategies are complementary. It recommended that premobilization training strategies should focus on crew qualification and platoon maneuver proficiency, while platoon gunnery and multiechelon training should be conducted as time allows (DAIG, 1991, pp. 1, 4-3). Existing postmobilization training plans had to be revised because they did not allow time to retrain and attain the prescribed standards or to accomplish administrative, logistics, and routine housekeeping tasks, including medical and dental screening (DAIG, 1991, p. 2-6). Soldiers missed critical individual and crew training to take care of medical problems, attend MOS courses, requisition supplies, or return to home station to retrieve equipment left behind. Personnel shortages, high crew turbulence, insufficient premobilization experience and training, and leadership problems contributed to crew proficiency shortfalls (DAIG, 1991, pp. 3-4, 3-6).

In congressional testimony, Army leaders drew a distinction between combat units and support units. For example, General Burba testified:

> Some of our most complex individual skills, such as medical as a good example, aviation is a good example, where there is civilian equivalency and which are easy to train on the weekend, are some of our most ready units. ... Those type skills, even though they are complex individually, are not complex in a collective sense ... . The tough ones are what we call our combat arms skills, cavalry, infantry and armor. ... Combat arms units must synchronize everything on the battlefield. ... It is an art. It is not a science, and it takes time to master. (U.S. House of Representatives, 1991a, p. 210)

Goldich, 1991, pp. 29–30, quotes Army Chief of Staff GEN Gordon Sullivan:

> The Forces Command (FORSCOM) Commander will specify that the *primary* focus of the combat units prior to mobilization be on individual soldier qualification and at the crew, squad, and platoon levels. Tank crews and platoons must be proficient because they are the building blocks for larger parent unit operations. When that is accomplished and as resources permit, higher level collective training can be conducted. ... Full scale company, battalion, and brigade operations will be the focus during postmobilization training. The training time available before call-up is insufficient to master the complex and highly perishable skills required at these levels. Training standards for small units are well defined—they are the standards of the Total Army—the mission training plans.

In a related provision, Title XI requires the Army to develop a system for identifying the priority for mobilization of RC units, based on regional contingency plans, that specifies the number of postmobilization training days each unit type will need. This system is to be linked to the resource allocation process so that units that have fewer postmobilization training days receive greater funding for training, full-time support, equipment, and manpower in excess of authorized strength. In 1992 congressional hearings, Representative William Dickinson stated: "We need to build on the lessons of Desert Storm and develop realistic expectations concerning Guard combat units, the time it takes to mobilize them, and how their capabilities fit into future U.S. force structures" (U.S. House of Representatives, 1992c, p. 424). On the same topic, the Army Inspector General, LTG Ronald Griffith, said:

> There are some single function units that can be expected to go very early. … [W]ith respect to medical units, there are some units that you could probably move within 72 hours. … [T]he amount of post-mobilization training depends in large measure on the type of unit.

Representative Les Aspin responded, "What we need to know is what is a realistic post-mobilization training schedule for various kinds of Guard and Reserve units?" (U.S. House of Representatives, 1992c, p. 440).

Congress was also concerned that the Army's readiness rating system gave a misleading picture of the time it would take to prepare the roundout brigades to deploy. Section 1121 of Title XI requires the Army to modify its readiness rating system for USAR and ARNG units to ensure that it provides an accurate assessment of the unit's deployability and any shortfalls that require additional resources. In particular, the personnel readiness rating is required to reflect

- the percentage of requirements that is manned and deployable
- the fill and deployability rate for critical occupational specialties
- the number of personnel who are qualified in their primary MOS.

The equipment readiness assessment is required to

- document all equipment required for deployment
- reflect only the equipment directly possessed by the unit
- specify to the effects of substitute items
- assess the effects of missing components on the readiness of major equipment items.

Goldich, 1991, p. 15, reports that all three roundout brigades were rated either C-2 (requiring 15 to 28 days of postmobilization training) or C-3 (requiring 29 to 42 days) at the time they were activated. The Army Inspector General found that expectations of initial levels of training and readiness of the roundout units were too high and that more-objective criteria and methods of measuring readiness were needed. These expectations were based on recent USRs, 1-R reports (prepared by AC division evaluators during annual training [AT]), and discussions with unit leaders. Shortfalls in crew proficiency were not clearly specified on 1-R reports. For example, most M-1 tank crews did not know how to boresight their tank weapons. One brigade that reported C-2 in personnel required over 600 replacements, had more than 500 non–MOS-qualified soldiers, and processed more than 300 personnel for release from active duty (DAIG, 1991, pp. 3, 2-3, 2-7, 2-8). The percentage of soldiers who were not qualified in their MOSs ranged from 11 to 21 percent in the brigades. More than 1,000 soldiers required MOS training, of whom about 65 percent were in combat arms MOSs. The Army Inspector General found that a lack of MOS specificity in AR 220-1 was the cause of false perceptions about MOS qualification, not false reporting (DAIG, 1991, p. 3-12).

All brigades mobilized with severe shortages in communications security equipment; radios; and nuclear, biological, and chemical defense equipment. Planners did not expect these shortages because the USR criteria exempted these items from the report (DAIG, 1991, p. 2-8).[7] To increase the accuracy of readiness reporting, the Army

---

[7]   Current readiness report criteria can be found in AR 220-1, 2010. The version of AR 220-1 that was in effect in 1990 was dated August 30, 1988, and can be found in the collection of the Pentagon Library. Although Army readiness reporting was changed in response to

Inspector General recommended that the Army improve the specificity of MOS qualification criteria in the USR and phase out non-reportable line items from equipment ratings. The Army Inspector General also recommended that the 1-R report should address a more-comprehensive range of the unit's mission-essential task list (METL), rather than focusing on training done at AT, and that it should be redesigned into two parts, the first addressing deficiencies that, if not corrected, would adversely affect postmobilization validation to deploy, and the second providing a definitive evaluation of AT (DAIG, 1991, pp. 4-6, 4-13, 4-14).

In congressional testimony, General Burba acknowledged,

> We need to formulate a new system whereby we evaluate the reserve units against two criteria—against what they can accomplish during those 39 training days a year and against their capability to fulfill general war reinforcing and contingency missions.

If you evaluate RC units to the same standards as AC units,

> you tend to have them training at too high a level, so because of only 39 days of training, they try to train at brigade, battalion, and lower levels. Therefore, they reach mediocre standards. ... If you evaluated them based on more realistic standards of what they could accomplish in 39 days, then they would go into postmobilization training at a much higher level. (U.S. House of Representatives, 1991a, p. 213)

Goldich, 1991, p. 38, cautions that

> it indicates a misunderstanding of the limits of the C-ratings to assume that Guard units having a particular C-rating should have

Title XI, it remains challenging to develop readiness metrics that fully reflect unit readiness for particular missions. See, for example, Pernin et al., 2013. The problems that report cites include subjective training ratings that are sometimes poorly correlated with personnel ratings, equipment ratings that do not reflect true equipping posture, a tendency to upgrade ratings as reports move up the chain of command, and an inability to link funding to changes in readiness ratings (pp. 82–88). The Commission on the National Guard and Reserves, 2008, p. 184, also expressed concern that "the existing readiness reporting system does not capture in adequate detail the readiness and capabilities of reserve component units."

had their deployment status determined solely by that C-rating. However, it was equally misleading, in the years before Desert Shield and Desert Storm, for both Guard and active Army leaders to overstate the actual readiness of the roundout brigades by pointing to their C-ratings, frequently as high as those of similar active Army units.[8]

Section 1120 of Title XI requires the Army to expand the use of simulations, simulators, and advanced training devices and technologies to increase training opportunities for ARNG units. This provision was intended to compensate for the small number of peacetime training days available to RC units and the limited availability of ranges and large maneuver areas, while also expanding opportunities for battalion- and brigade-level commander and staff training. General Burba testified to Congress: "Related to the difficulty of mastering complex skills in short training periods is the fact that, in many cases, units are not within geographic reach of the facilities and large maneuver areas where these skills can be practiced" (U.S. House of Representatives, 1991a, p. 172). COL Fletcher Coker, commander of the 155th Armored Brigade in Mississippi, stated that, "ideally, a platoon is just about the biggest unit that you can freely maneuver on Camp Shelby as presently configured. … We need wider ranges, more complex ranges, and more maneuver space" (U.S. House of Representatives, 1991a, p. 202).

To improve the readiness of the roundout brigades, the Army Inspector General report recommended that the Army improve training facilities, ranges, and training aids, devices, simulators, and simulations (TADSS); provide annual battle staff training, reinforced with exportable training packages; and develop self-help training support

---

[8] Goldich, 1991, found a range of opinions among Army and ARNG leaders about the amount of postmobilization training time the roundout brigades would need, but all agreed that some postmobilization training would be required. The roundout brigades were not intended to be used as contingency forces for immediate, short-duration deployments. However, he found that

> both active Army and Army National Guard leadership left the impression, in [both] public comments and congressional testimony, that the roundout brigades would and could deploy with their parent divisions under all circumstances, without any explicit reference to the time that might elapse between mobilization and deployment. (p. 19)

packages for roundout units, including TADSS, that would be pre-positioned at mobilization stations (DAIG, 1991, pp. 4-6, 4-9, 4-16). Regarding simulation training, General Burba testified:

> Battalion and brigade level commanders and staff training should be focused on simulation training. It must be aggressively pursued so that our leaders can be trained to orchestrate the complex operating systems of today's airland battlefield. This includes attendance at the Tactical Command Development Course and frequent use of battle simulations. (U.S. House of Representatives, 1991a, p. 175)

MG John D'Araujo, Director of the ARNG, stated that,

> in the pre-mobilization phase, we need training programs that will allow higher levels of tank gunnery sustainment. For example, we may need to require extensive use of simulators, much as we do with our aviation force. … [W]hen we say we need to train the tank crews on simulators, we have got to make sure those things are there. (U.S. House of Representatives, 1991a, pp. 214–215)

Several provisions of Title XI were intended to improve the personnel readiness of ARNG units. Section 1111 sets a goal of increasing the percentage of ARNG personnel with prior AC experience to 65 percent for officers and 50 percent for enlisted by the end of FY 1997. The Army Inspector General had found low premobilization manning levels in critical combat arms and low-density support specialties and MOS qualification shortfalls in the roundout brigades (DAIG, 1991, p. 2). The report also found that poor leadership, especially in the NCO ranks and among field-grade officers, hindered training. Many leaders could not recognize standards or hold their soldiers to recognized standards. They lacked sufficient opportunity to train leadership skills during the premobilization period and had not attended available leader development courses. To address these problems, the report recommended that the Army increase MOS qualification rates through increased recruitment of prior-service soldiers in critical skills

(DAIG, 1991, pp. 3-7, 4-7). Congress saw the Army's planned draw-down as an opportunity to increase the proportion of ARNG soldiers with AC experience. Representative John Spratt stated that, with "the Active Army beginning to downsize, there should be quite a few NCOs or well-trained specialists who are coming out of active duty who might be recruited by the Guard" (U.S. House of Representatives, 1991a, p. 220).

However, these goals quickly proved to be challenging to meet. In 1993 congressional hearings on the implementation of Title XI, William Clark, acting Assistant Secretary of the Army for Manpower and Reserve Affairs (ASA[M&RA]) testified:

> The percentage of prior service personnel currently in the Army National Guard is 52 percent for officers, 78 percent for warrant officers, and 48 percent for enlisted members. (U.S. House of Representatives, 1993b, p. 173)

> Increasing the proportion of National Guard officers with Active experience to 65 percent and the enlisted force to 50 percent is another complex and also potentially costly program. ... We will have to increase the propensity of soldiers departing Active Duty to enlist and to remain in the Guard and Reserve. We may also have to rotate more [Guard] officers through 2- and 3-year stints on Active Duty to meet the requirements of the law. (U.S. House of Representatives, 1993b, p. 165)

> [W]e are currently enlisting in the Guard and the Reserve ... about 50 percent prior service people today. ... [T]he challenge is going to be not in bringing them in but in making sure you retain them with the Guard units. (U.S. House of Representatives, 1993b, p. 180)

Section 1114 mandates that military education requirements for NCOs be met before promotion to a higher grade and that the Army ensure that sufficient training positions are available to meet these requirements. Reports and congressional testimony indicated that both NCOs and commissioned officers in the roundout brigades lacked required training courses. Richard Davis, Director of the Army Issues Group at GAO, testified to Congress:

> NCO leadership skills were lacking. For example, in one of the roundout brigades, only 30 percent of the NCOs that were required to have the basic NCO course actually had that course. ... Commissioned officers, much the same story. ... Again, in one of the roundout brigades, ... only 27 percent of the officers required to have an advance officer course actually had taken one. (U.S. House of Representatives, 1992c, p. 437)

The Army Inspector General reported that, prior to ODS, only AC commanders had attended the Tactical Commanders' Development Course. Commanders and staffs from the roundout brigades missed some collective training to attend this course. The report recommended that the Army require and fund increased officer and NCO education and train-the-trainer courses and enforce NCO Education System completion as a prerequisite for NCO promotions (DAIG, 1991, pp. 3-6, 4-5, 4-10).

Goldich, 1991, pp. 21, 31, reported:

> Individual officers and soldiers were either not capable of performing, or in many cases were not even aware of the range of, tasks they had to perform as part of a combat unit in the field, as opposed to the part-time environment in which they had been soldiering before mobilization. ... A large number of officers and NCOs had to be removed from their units and sent to formal school courses after mobilization. ... [T]his removed them from their units precisely when those units were themselves training to meet deployment standards, creating further leadership and training problems that took more time and effort to resolve. ... [H]ad their officers and NCOs had more opportunity for, and/ or requirements to, attend active Army schools to obtain necessary technical, tactical, and leadership training, then fewer deficiencies would have needed to be remedied after mobilization. ... [T]hese requirements [were] frequently waived or honored more in the breach than in fact.

Section 1115 of Title XI requires the Army to establish a personnel accounting category for ARNG members who have not completed the minimum training required for deployment or who are otherwise not available for deployment. Section 1116 requires the Army to transfer members who do not meet the minimum physical standards for deployment to this category within 90 days.[9] The Army Inspector General cited problems with both vacancies and nondeployable soldiers in the roundout brigades. The reserve call-up authority used for the ARNG brigades did not allow the Army to activate Individual Ready Reserve members to fill vacancies. Personnel turnover caused by instant promotions and cross-leveling were major reasons for instability of crews and degraded training. After cross-leveling, many units still had shortages in critical combat arms and low-density CSS specialties, and these skill shortages were not corrected by the time the units were mobilized. Many non–line-of-duty problems were not expeditiously handled by medical evaluation boards, and the affected personnel remained with the unit at call-up. Soldiers found to be nondeployable during the postmobilization process further degraded units already at less than 100-percent personnel fill (DAIG, 1991, pp. 3, 3-10, 4-6, 4-17). The Army Inspector General's report recommended that the Army authorize and require personnel fill up to 110 percent of authorized strength in key combat crew and low-density MOSs (DAIG, 1991, p. 4-7).

General Burba testified to Congress in 1991: "[I]n any Reserve unit there are always some soldiers who are not deployable. This includes soldiers who have not yet undergone their initial active duty training, are still in high school, or have a temporary condition making them nondeployable" (U.S. House of Representatives, 1991a, p. 220). In 1993, William Clark, acting ASA(M&RA), stated

---

[9] Policies regarding trainees, transients, holdees, and students (TTHS) accounts for the Army Reserve and National Guard have changed over time. As of 2010, the AC TTHS account was authorized 71,000 personnel (13 percent of end strength); the ARNG TTHS account was authorized 8,000 personnel (2.5 percent); and the USAR TTHS account was authorized 4,000 personnel (2 percent). At that time, the ARNG used only the "trainee" portion of the account to cover part of its total training pipeline of 29,000 soldiers. The USAR only used the "holdee" portion of the account for soldiers who were medically nondeployable (White, 2010, pp. 15–16).

> We need more rigor in identifying and temporarily moving non-deployables out of the unit strength. It gives a better, more accurate portrayal of what the organization's capability really is. Also, it then puts the demand on the individual to correct that particular condition. (U.S. House of Representatives, 1993a, p. 160)

Section 1117 required annual medical and dental screening of all ARNG members and a full physical examination every two years for members over 40, while section 1118 required the Army to develop a plan to ensure that ARNG units scheduled for early deployment were dentally ready. The Army Inspector General reported that postmobilization medical nondeployable rates ranged from 5 to 10 percent of assigned strength. Dentists categorized 30 to 35 percent of all soldiers as dentally unfit for deployment (Dental Class 3 or 4) due to untreated dental problems or poor-quality panographic X-rays.[10] Nearly 15 percent of personnel assigned to the brigades required but lacked complete over-40 medical examinations and cardiovascular screenings (DAIG, 1991, pp. 3-12, 3-13). The Army Inspector General recommended that the Army require over-40 medical examinations, conduct an annual 100-percent screening of medical and dental records, provide resources to maintain dental readiness, and expedite medical board processing for soldiers with permanent medical profiles (DAIG, 1991, pp. 4-7, 4-8) In congressional hearings, Representative Beverly Byron stated:

> I noted time and time again the question on the physical capability, the critical dental problems, 50 percent failed initial physical tests. … [I]t is an issue that the Guard, if they are going to be credible, needs to put more of an emphasis on. (U.S. House of Representatives, 1991a, p. 222)

---

[10] At the time, premobilization dental care could only be provided during AT, which might be considered an unacceptable training distractor (DAIG, 1991, p. 3-12).

However, these provisions proved to be expensive to implement and were repealed in the NDAA for FY 1996 (PL 104-106, 1996, section 704).[11] William Clark testified to Congress in 1993:

> We are looking very carefully at using both Active Army, Army Reserve, and Army National Guard dental resources. ... We did have a proposal ... that involved taking care of the early deploying units. It did have a relatively high price tag. We are trying to look at other alternatives so we can minimize those costs and then apply this to those units in which there is indeed a requirement for them to be prepared ... right at the very beginning. ... We may have to look at accession standards too. (U.S. House of Representatives, 1993b, pp. 182–183)

Section 1134 of Title XI requires the Army to report annually to Congress on the compatibility of AC and RC equipment, the effect of equipment incompatibility on combat effectiveness, and a plan to achieve full equipment compatibility. One of the roundout brigades had only recently received its Abrams tanks and Bradley fighting vehicles and had not completed new equipment training when it was mobilized. BG Gary Whipple, commander of the 256th Infantry Brigade, Louisiana National Guard, testified to Congress:

> The 256th Brigade received the Abrams tank in 1989, and in the summer of 1990 the brigade received our Bradley fighting vehicles. In each case the first training phase allowed us only enough time to learn to drive and the basics of maintenance of the equipment. We had not had the opportunity to train to shoot the weapons systems or tactically maneuver them on the ground. ... [N]ot all of the brigade's equipment matched that belonging to the active 5th Infantry Division units, and we also had an equipment readiness problem with equipment we were authorized, but had not yet received, or for which we had inadequate substitutes. (U.S. House of Representatives, 1991a, pp. 188–189)

---

[11] Section 704 of the NDAA for FY 1996 added provisions requiring medical and dental screening to 10 USC 1074d, but they only apply to Selected Reserve units scheduled to deploy within 75 days after mobilization.

A related problem was that many operators and mechanics were not adequately trained to maintain the brigades' equipment. General Burba testified that,

> on a day-to-day basis, reserve component unit equipment is reasonably well maintained. Only when a unit moves to the field for an extended period of time does it become apparent that operator knowledge, mechanic diagnostic skills, and knowledge of the Army maintenance system are generally lacking. (U.S. House of Representatives, 1991a, p. 171)

The Army Inspector General reported that maintenance problems surfaced quickly because of shortfalls in operator, crew, and organizational maintenance training; overdependence on full-time maintenance personnel, and shortages of organization and direct support mechanics. Many operators and crews could not perform preventive maintenance checks and services. Crews and mechanics lacked opportunities for realistic hands-on maintenance training because equipment was stored and maintained at centralized facilities, not at home station (DAIG, 1991, pp. 3-14, 3-15). The Army Inspector General recommended that better integration of roundout units into the AC maintenance system was needed to train RC mechanics, operators, and logistics clerks and that full-time maintenance technicians should be assigned to the units they support. In practice, they tended to be assigned to DS- and GS-level units that provided more promotion opportunities (DAIG, 1991, pp. 2, 4-8).

Section 1133 requires the Army to develop and implement a program to ensure that automated systems for personnel, supply, maintenance management, and finance are compatible across all Army components. The Army Inspector General reported that all three roundout brigades required too much time and effort to transition from their premobilization systems and procedures to AC systems and procedures, particularly in personnel and logistics. Severe personnel and pay problems resulted from lack of interface between AC and RC automated systems and from lack of RC personnel trained on AC systems. One brigade took more than 90 days to transfer its personnel records to the AC system because its supporting unit had deployed. The brigade's transition to AC logistics systems was slow and incomplete. Unit supply personnel expended significant time and effort ordering authorized supplies and frequently encountered problems properly submitting requisitions for repair parts. Supply clerks rarely performed MOS-related duties at inactive duty training (i.e., weekend drills) and relied on full-time personnel during AT (DAIG, 1991, pp. 3, 3-9 to 3-15). The Army Inspector General recommended that the Army conduct a combined AC-RC review of systems to eliminate differences and establish commonality of equipment, terms, procedures, and data field structures (DAIG, 1991, p. 4-16).

However, Congress put relatively few of these provisions into the U.S. Code itself. One of the exceptions is in 10 USC 12001, which requires the Army to

> carry out a program to provide active component advisers to combat units, combat support units, and combat service support units in the Selected Reserve of the Ready Reserve that have a high priority for deployment. … The advisers shall be assigned to full-time duty in connection with organizing, administering, recruiting, instructing, or training such units.

The section sets the number of AC personnel assigned as advisers at 3,500 and allows the Army to count those who are assigned to an AC unit whose primary mission is to provide dedicated training support to RC units. It also requires the Army to provide an annual report on implementation of the program as part of the Army Posture Statement (10 USC 12001). Most of the items that must be included in this report are listed in 10 USC 10542 and can be directly linked to the provisions of Title XI.

Implementation of Title XI and Subsequent Changes in the 1990s

The Army's implementation of Title XI was integrated with Bold Shift, the pilot program that the NDAA for FY 1992 and 1993 established. William Clark, acting ASA(M&RA), described the program as follows:

Bold Shift … consists of several high pay-off training and readiness programs. These include Operational Readiness Evaluations, and increased Active component training support of Guard Round Out and Round Up units, and other Guard and Army Reserve early deploying units in the Contingency Force Pool. Other Bold Shift initiatives are aimed at strengthened command and staff proficiency from company through division level, increased involvement of the wartime chain of command, leader development, and establishing new organizations for the training support of combat units. The pilot program adding 2,000 additional Active component soldiers to train the Reserve components is also managed under Bold Shift. (U.S. House of Representatives, 1993b, p. 166)

Bold Shift refocused collective training for RC combat units at the crew, squad, and platoon levels, as required by Title XI. The program adopted a new style of training called the Reserve Training Concept, which centered on highly structured and supported training on selected tasks in training events called "lanes." AC advisers and AC associate units provided the opposing force, observer controllers, crew examiners, and other training support personnel so that the RC unit could focus on training and increase the number of tasks trained. An AAR covering the unit's performance and discussing areas for improvement followed each training event. The unit had to perform each training event to published Army standards before proceeding to the next training lane (Sortor et al., 1994, pp. 8–11).

The Military Forces and Personnel Subcommittee of the House Committee on Armed Services conducted two hearings on AC support of RC training and implementation of Title XI in April 1993. Clark described how the AC advisers were being assigned: "There are three principal elements that we have in this program. The first is the Resident Training Detachment. That's similar to the Marine Corps I&I [Inspector/Instructor] program." They "live and work with the sponsored RC unit" and

primarily support Army National Guard Roundout and Roundup combat units from brigade through company level. The Regional Training Teams are located to provide training assistance to the smaller combat support and combat service support units on a regional basis. ... Operational Readiness Exercise teams operate on a regional basis to provide an assessment of unit training and unit readiness on a single standard for all three components. (U.S. House of Representatives, 1993a, pp. 131–132)

AC associate units also provided RC training support under Bold Shift. Clark stated: "[O]ur Roundout and Roundup brigades ... are associated with specific divisions. ... [T]he division has direct responsibility for its training." LTG John Tilelli, Deputy Chief of Staff for Operations and Plans, commented:

> Having come out of division command a little over 8 months ago, I would send to individual training weekends a mobile training team to assist in the training of the units. ... [W]hen we did the Active Duty 2 week period, ... I had an Active component brigade ... do all of the opposing forces work, the observer controller work, setting up the training lanes, and assisting in the instruction for the command, staff and unit training. This would be on top of the Resident Training Detachments [associated with the roundup brigade]. (U.S. House of Representatives, 1993a, pp. 149–150)

The Army's statement to the subcommittee indicated that it was in compliance with 11 of the 18 sections of Title XI but was concerned about the costs and resource implications of some of the remaining provisions. Clark testified:

> The Army is continuing to develop options for meeting the medical and dental readiness requirements. ... The National Guard Bureau is developing a Non Deployable Personnel Account and system for tracking individual soldiers. ... Of particular concern is the requirement for an additional 3,000 Active component advisers. ... [I]t could take the leadership of the current 1st Infantry, 2nd Armor, and 4th Infantry Divisions ... or nearly one half of the instructors at our TRADOC [U.S. Army Training and Doctrine Command] schools to meet this requirement. ... The Army will need flexibility with implementation dates and resources. (U.S. House of Representatives, 1993b, pp. 167–168)

MG James Lyle, Director of Training, Office of the Deputy Chief of Staff for Operations, added, "It is resource intensive because the kind of people you want out there are sergeant first class and above in the NCO corps, warrant officers and essentially captains and above on the officer side" (U.S. House of Representatives, 1993b, p. 176). Congress later delayed the implementation date for the increase in AC advisers from the end of FY 1994 to the end of FY 1996 (PL 103-337, 1994).

Evaluations of Bold Shift

Sortor et al., 1994, evaluated the initial effects of Bold Shift in the first year of its implementation, based on observations of AT; data collected from the Training Assessment Model, Operational Readiness Evaluations, and other sources; and surveys of personnel in RC units that participated in Bold Shift. They found that the main features of the program (training to more realistically attainable premobilization goals, the Reserve Training Concept, and closer ties between the AC and RC) seemed to be moving in the right direction and were well worth continuing. A large majority of the unit members they surveyed thought Bold Shift was effective in improving the readiness of their units for their wartime missions and felt that the program should continue.

However, the RC units participating in the pilot program were not able to meet their premobilization training and readiness goals. Specifically, only about one-third of CS and CSS units were able to meet their proficiency goals in critical tasks to the company level. In combat units, less than 30 percent of crews qualified on Table VIII, the final crew-level gunnery qualification exercise for M1 tanks and Bradley fighting vehicles. Even fewer executed maneuver training lanes at the platoon level:

> Most brigades had to choose, with limited time, between focusing on gunnery and maneuver or sending individuals to school for MOS qualification and other individual training. There was simply not enough time in their schedules to practice all of the tasks they were expected to master. (Sortor et al., 1994, p. xv)

One of the most difficult challenges units participating in Bold Shift faced was attendance at AT: Only 60 to 70 percent of members attended AT with their units. As a result, only about 68 percent of authorized M1 crews and 50 percent of Bradley crews were present at AT. Many of the remainder were attending individual training courses, including MOS qualification courses and required professional training for NCOs. Many NCOs had not received required training. For pay grades E5 and above, 15 to 25 percent of soldiers were not qualified in their duty MOSs; 37 percent of E5s needed to take the Primary Leadership Development Course; 39 percent of E6s needed to take the Basic NCO Course; and 29 percent of E7s needed to take the Advanced NCO Course. Only one-half of company commanders had attended the Officer Advanced Course, and 39 percent of battalion commanders had completed the precommand course (Sortor et al., 1994, pp. xvi, 19, 37–38, 59).

The House Subcommittees on Military Readiness and Military Personnel asked GAO to evaluate Bold Shift in 1994, focusing on whether it had enabled combat brigades to meet peacetime training goals, whether AC advisers were working effectively to improve training readiness, and whether the brigades would be ready to deploy 90 days after mobilization (GAO, 1995). GAO examined the training results of the seven former roundout brigades, which had been redesignated as enhanced brigades following the Bottom-Up Review in 1993.[12] Their analysis was based on discussions with brigade commanders and other officials, AC advisers, and representatives from three affiliated AC divisions; data collected from the brigades and the Army; and models the Director of Army Training, Army Inspector General, and RAND Corporation had developed to estimate the amount of postmobilization training the brigades would need (GAO, 1995, Ch. 1).

GAO found that none of the seven brigades had come close to meeting the Bold Shift training goals. In 1993, the combat platoons were fully trained in an average of only 14 percent of their METLs. Tank and Bradley crews in only four of 13 battalions met gunnery

---

[12] There were a total of 15 enhanced brigades, including mechanized infantry, armor, armored cavalry, and light infantry units. They were responsible for reinforcing AC units if the number of AC units was insufficient to resolve two nearly simultaneous major regional contingencies (GAO, 1995, Ch. 1).

goals.[13] Twelve of 18 battalions were able to meet gunnery goals in 1994, but their heavy focus on gunnery left little time for maneuver training. Another brigade focused so heavily on METL training that it did not even attempt to qualify on Table VIII in gunnery. Although three of the brigades met the goal that 85 percent of soldiers would be qualified in their duty MOSs, only about 70 percent of officers and 58 percent of NCOs had competed required military education courses (GAO, 1995, Ch. 2).

GAO also found that the role of the AC advisers was not well defined. Army guidance was not clear about whether they were supposed to identify and resolve training problems in the ARNG brigades or were only to assist with training. When advisers did attempt to correct training problems, the ARNG units were not always responsive to their suggestions. According to some AC officials, the effectiveness of the advisers depended on the quality of their working relationship with the brigades. The GAO was pessimistic that the enhanced brigades could be ready to deploy within 90 days after mobilization because the postmobilization training models it examined generally assumed that the brigades would be able to meet the Bold Shift premobilization training goals (GAO, 1995, Results in Brief).[14]

Training Support XXI

By 1996, RC training support organizations had been consolidated into regional training brigades under the CONUSAs, which supported combat arms units, and divisions (exercise) under USARC, which supported CS and CSS units. This structure reduced the role of AC corps and divisions in training RC associate units and increased the role of the CONUSAs. Arnold, 2003, argues that it was an improvement over previous organizational structures, but was challenging from the perspective of RC commanders. A typical unit had to coordinate with

---

[13] The goals were to qualify 60 to 66 percent of Bradley crews, depending on the vehicle model, at the gunnery Table VIII level. Tank battalions were expected to qualify 75 percent of their assigned crews.

[14] However, note that the 48th Brigade was validated as ready for deployment to ODS about 90 days after activation, without the benefits of Bold Shift.

up to four different organizations to plan training and could not rely on having the same units provide support from one training year to the next. Headquarters, Department of the Army (HQDA) directed FORSCOM to conduct a functional area assessment of AC/RC training support organizations. The goals were to assess the structure of these organizations and to recommend ways to eliminate redundancies and improve unity of command for training support units. The assessment recommended a tricomponent organization with a single chain of command and a single point of contact for RC commanders to coordinate training. The new structure was also intended to increase coordination between premobilization and postmobilization training and further reduce the roles of AC corps and divisions by giving the CONUSAs responsibility for both pre- and postmobilization training and mobilization assistance (Arnold, 2003, pp. 3–5).

The Army began implementing Training Support XXI in 1997, and the new training support organizations were fully implemented on October 1, 1999. The regional training brigades and divisions (exercise) were reorganized into TSBs, which came under training support divisions (TSDs) under the operational control of the CONUSAs. There were five TSDs, three in First Army and two in Fifth Army, and one additional TSB under the command and control of U.S. Army Pacific to provide support to RC units in the U.S. Pacific Command area of responsibility. Each TSD had a specific area of responsibility in which it coordinated, synchronized, and supervised training support.

Each TSD comprised several TSBs, as determined by the training requirements in the area of responsibility. Each TSB had training support battalions (TSBns) organized to support units with lane training, evaluations, and observer controller-trainers. Combat arms TSBns were staffed with AC personnel, while CS/CSS TSBns had both AC and RC soldiers assigned. Each TSB also had a logistics battalion staffed entirely with RC personnel who maintained the TSB's equipment and vehicles (Arnold, 2003, pp. 7–9). However, training support resources were focused on units that had a high priority for mobilization; others received little or no attention. TSBs assisted priority units in developing their METLs, yearly training briefs, and plans for weekend drill and AT. TSB representatives attended unit drills, provided input on training plans and performance, and provided formal evaluations each year as part of the supported unit's AT (Chapman, 2008, p. 6).

According to Chapman (2008, p. 6), Training Support XXI had a positive impact on the mobilization of RC units for operations in Iraq and Afghanistan:

> After the 9/11 attacks, the TSBs were able to rapidly shift their focus to the planning and execution of postmobilization training for mobilizing reserve component units, to include validating that RC units are ready to complete the missions for which they have been mobilized. Were it not for the TSBs, other AC units would have had to assume this function in addition to their other missions, to the detriment of both RC unit postmobilization training and the supporting AC units' own deployment preparations.

CHAPTER THREE

Evolution of Pre- and Postmobilization Training to Support Operations in Iraq and Afghanistan

This chapter describes additional changes in pre- and postmobilization training made to support the rotational demand for forces in OEF and OIF and provides some evidence on the number of pre- and postmobilization training days needed to prepare various types of RC units for deployment. This analysis also examines the effects of the Secretary of Defense's 2007 memorandum limiting total mobilization time to 12 months (Gates, 2007), which caused the Army to shift some training to the premobilization period to maximize the time that units could spend in the theater of operations. The chapter concludes with RC unit perspectives on postmobilization training obtained through First Army's AAR process.

Changes to RC Training Support

At the time of the terrorist attacks on the United States in 2001, First Army was responsible for training support of RC units located east of the Mississippi River, while Fifth Army supported training of RC units west of the Mississippi River. As the number of RC units mobilized to support operations in Afghanistan and Iraq increased, the roles of the CONUSAs shifted to supporting postmobilization training of RC units preparing to deploy. Initially, SRP and much of the individual soldier qualifications and training, as well as collective training, were done after mobilization. RC units that deployed early in OEF and OIF had relatively short notice to prepare for deployment, but as notification periods increased, personnel turnover in RC units preparing to deploy still made it difficult to schedule required training more than a few months before mobilization.[1]

In the mid-2000s, the Army began to reorganize itself in two respects. First, it converted from a division-based force to a more-flexible, modular force based on brigade combat teams (BCTs) and multifunctional and functional support brigades with standard-ized organizational designs for the AC and RC. Second, it converted from a tiered readiness model, under which higher-priority units received more resources, to a cyclical readiness model, known as the ARFORGEN process. Under ARFORGEN, units cycle through three force pools—reset, train/ready, and available—building their train-ing readiness and capabilities over time until their available year, when they may deploy to meet an operational requirement or be available for contingency requirements that arise during that year. The purpose of the ARFORGEN process is to provide a sustained flow of trained and ready forces, while establishing a more predictable deployment tempo for soldiers and families (AR 525-29, 2011). Also as part of this reorga-nization, First Army was given responsibility for supporting pre- and postmobilization training of all U.S.-based RC units in 2006,[2] while its mission to provide military support to civil authorities in the eastern United States was consolidated under Fifth Army, which became U.S. Army North, the Army component of U.S. Northern Command (First Army, undated).[3]

---

[1]   Lippiatt and Polich (2010) found that only 50 to 60 percent of soldiers who deployed with RC units had been assigned to the unit for at least one year before deployment. For example, if a typical ARNG BCT held a premobilization training event seven months before deploy-ment, about 35 percent of the soldiers who eventually joined the unit and deployed with it would have missed the event. AC units had similar rates of personnel turnover, but full-time soldiers have more training time available than RC soldiers do to complete training require-ments they may have missed.

[2]   As of 2012, First Army had two divisions (Division East at Fort Meade, Maryland, and Division West at Fort Hood, Texas) and 16 TSBs located at nine mobilization force genera-tion installations (First Army, 2013a).

[3]   U.S. Northern Command was established in October 2002 to provide command and control of DoD homeland defense efforts and to coordinate defense support of civil authori-ties (U.S. Northern Command, 2013).

In January 2007, Secretary of Defense Robert Gates issued a memorandum limiting involuntary mobilizations to a maximum length of one year, excluding individual skill training and postmobilization leave at the services' discretion. The memorandum also set a goal of one year mobilized to five years demobilized for RC units, recognizing that the services might not be able to meet this goal immediately (Gates, 2007). Since the Army had been deploying RC units for a year (or sometimes more) after postmobilization training, this policy required the Army to rebalance pre- and postmobilization training to maximize the amount of time RC units could spend with "boots on the ground" in the theater of operations.

In response to this policy change, the Deputy Chief of Staff, G-3/5/7, published Execution Order (EXORD) 150-08 in February 2008 (Thurman, 2008). The purpose of EXORD 150-08 was to define pre- and postmobilization training roles and responsibilities; premobilization training tasks, including documentation requirements and standards; and an integrated deployment training process, from notification of sourcing to deployment, supported by frequent assessments to allow adjustments to training plans and resource requirements. EXORD 150-08 defined First Army's responsibilities to include

- execute training and readiness oversight authorities (as defined in Joint Publication 1-02)[4] over RC forces assigned to the combatant commander

---

[4] Joint Publication 1-02, 2010 [2013], p. 296, defines these as

> The authority that combatant commanders may exercise over assigned Reserve Component forces when not on active duty or when on active duty for training. As a matter of Department of Defense policy, this authority includes: a. Providing guidance to Service component commanders on operational requirements and priorities to be addressed in Military Department training and readiness programs; b. Commenting on Service component program recommendations and budget requests; c. Coordinating and approving participation by assigned Reserve Component forces in joint exercises and other joint training when on active duty for training or performing inactive duty for training; d. Obtaining and reviewing readiness and inspection reports on assigned Reserve Component forces; and e. Coordinating and reviewing mobilization plans (including postmobilization training activities and deployability validation procedures) developed for assigned Reserve Component forces.

- publish guidance that
  - describes procedures for how it will exercise premobilization training and readiness oversight
  - identifies deployment tasks to be trained during pre- and post-mobilization
  - establishes clear training standards and documentation procedures
- establish in-process reviews (IPRs) and readiness reporting requirements, in addition to USRs
- provide training support and enablers for premobilization training, as requested by RC unit commanders
- request training support and enablers for postmobilization training from the reserve components
- validate RC units for deployment
- in coordination with RC commanders, develop and adjust deployed METLs and deployment training plans and determine capability levels at mobilization date
- develop deployment training plans for individual soldiers and units that have not completed required training by the mobilization date or that arrive after unit mobilization
- report RC unit readiness to HQDA from mobilization date to latest arrival date in theater.

EXORD 150-08 required an initial IPR no later than 60 days after notification of sourcing (usually two years before mobilization), at which the RC commander and First Army reviewed the unit's deployed METL and deployment training plan. The order required a joint assessment IPR 180 days before mobilization, at which First Army and the RC chain of command would

- confirm task completion
- forecast the projected completion date for remaining premobilization training tasks and the readiness level of the RC unit at mobilization
- forecast tasks to be completed after mobilization
- estimate the number of days required for postmobilization activities.

Depending on unit capability projections at the mobilization date, First Army was required to make necessary adjustments to the mobilization date and develop a postmobilization resourcing and training support plan. The premobilization training focus for the RC unit was to complete individual training and readiness activities, including medical and dental, and to conduct collective training to the maximum extent possible. EXORD 150-08 also specified an optional IPR 60 days before mobilization to provide a final assessment and allow adjustment of the deployment training plan to complete remaining premobilization requirements and adjust postmobilization training. The first general officer in the RC unit's chain of command was given authority to approve the RC commander's certification of deployment.

Since 2007, the ARNG and USAR have also increased premobilization training support for RC units, primarily using overseas contingency operations funding. The NGB authorized each state and territory to create premobilization training assistance elements (PTAEs) consisting of a three-person command-and-control cell and training assistance personnel in a ratio of one trainer for every 60 soldiers in units preparing to deploy. PTAE personnel are ARNG members recently returned from a deployment who volunteer to remain on active duty for up to two years. They are trained by First Army as observer controller-trainers and should also attend the Total Army Instructor Trainer Course and Small Group Instructor Course. PTAE training assistance personnel ensure that the mobilizing unit's training is conducted to standard and is properly documented (Weiss, 2008, and Sanzo, 2008).

The USAR established three regional training centers (RTCs), at Fort McCoy, Wisconsin; Fort Hunter-Liggett, California; and Joint Base McGuire-Dix-Lakehurst, New Jersey. The RTCs were staffed by mobilized soldiers and primarily conducted premobilization individual training on theater-specific individual readiness tasks,[5] Army warrior

---

[5] Theater-specific individual readiness training includes computer-based training and briefings on such topics as antiterrorism, operational security, injury prevention, suicide prevention, cultural awareness, Army core values, rules of engagement, first aid, and casualty evacuation. See Camp Atterbury, 2014a and 2014b.

training skills,[6] and weapon qualifications. The RTCs also provided some collective training, such as live-fire exercises, military operations on urbanized terrain, search operations, and convoy training. USAR units preparing for deployment typically scheduled an RTC rotation two to three months before their mobilization date (Schuette, 2008, and Flores, 2012).

Analysis of Pre- and Postmobilization Training Days

To examine the number of premobilization and postmobilization training days various types of RC units need to prepare for deployment, we used a database RAND developed in collaboration with the Defense Manpower Data Center (DMDC). This database has been used in several previous studies, including Lippiatt and Polich, 2010, and Lippiatt and Polich, 2013. It includes data from four sources:

- **individual personnel history:** pay grade, occupational specialty, entry date, initial military training, unit assignment, and other characteristics from DMDC's Work Experience File
- **activation and deployment:** month of activation and return from active duty and month of deployment to theater and redeployment to the United States, from DMDC's Defense Mobilization and Deployment database
- **pay:** records of actual pay, allowances, bonuses, and other monetary compensation (including hostile-fire pay), from the Reserve and Active Duty Pay Files
- **authorizations:** Army Master Force files describing unit organization and structure.

---

[6] Army Warrior Tasks are individual-level skills organized in five areas: shoot (maintain, employ, and engage with assigned weapon system; employ hand grenades), move (perform individual movement techniques, navigate from one point to another, move under fire), communicate (perform voice communications, use visual signaling techniques), survive (react to chemical or biological attack, perform immediate lifesaving measures; perform counter–improvised explosive device [IED], maintain situational awareness, perform combatives), and adapt (assess and respond to threats, adapt to changing operational environments; grow professionally and personally). See Headquarters, Department of the Army, 2014.

Using this database, we were able to identify the mobilization and deployment dates of the units and the average number of training days for the soldiers who deployed with the units in the year before mobilization and in the period between mobilization and deployment.[7] We define premobilization training time as the average of inactive duty training plus AT days for soldiers who deployed with the unit, excluding some individuals who might skew the results, such as AGR, other full-time support personnel, unqualified soldiers, soldiers who were assigned to deploying units but did not deploy with their units, and any training prior to the soldier's assignment to a deploying unit. We define postmobilization preparation time as the time between mobilization and arrival in theater, which includes time spent at home station; movement to the mobilization station; reception, staging, onward movement, and integration; training days; leave; load-out; and deployment. Since 2008, nontraining time in the period has been about 12 to 15 days. We divided the analysis into two periods, 2003–2007 and 2008–2010, to examine the effects of the 12-month limit on mobilizations.

The sample of RC units included in our analysis is shown in Table 3.1. It consists of 45 BCTs and 626 enabler units, including battalion headquarters, companies, and detachments, that deployed in 2003 through 2010. We grouped BCTs based on their mission type, such as counterinsurgency, security force, or training and assistance of Afghan or Iraqi forces.[8] We grouped enabler units based on unit type and likely deployment category, based primarily on doctrinal descriptions of unit employment. Category 3 is defined as units that travel or conduct most of their mission off forward operating bases (FOBs), category 2 has some travel off FOBs, and category 1 rarely, if ever, travels off FOBs.

Figure 3.1 shows the number of days needed for pre- and postmobilization training and preparation for BCTs, by mission type and

---

[7] However, these databases did not allow us to observe any additional in-theater training that may have occurred after deployment.

[8] Note that none of the BCTs trained for combined arms maneuver missions, which could add to the total amount of predeployment training and preparation time needed.

Table 3.1
Reserve Component Units Selected for Analysis

Category	Number	Mission/Type
BCTs	9	Counterinsurgency
	27	Security force
	9	Train and assist
Enabler units		
Battalion headquarters	52	Headquarters and headquarters company, combat sustainment support, and corps support battalions
	19	Headquarters and headquarters detachment, engineer battalions
Likely category 3 companies[a]	35	Engineer combat support companies
	99	Military police combat support companies
	157	Truck companies
Likely category 2 companies[b]	47	Engineer construction companies
	37	Maintenance companies
	22	Medical area support companies
Likely category 1 companies and detachments[c]	41	Adjutant general battalion headquarters and detachments
	46	Financial management detachments
	29	Public affairs detachments (mobile)
	42	Quartermaster field services and supply companies

[a] Travel or conduct most of mission off FOB.
[b] Some travel off FOB.
[c] Rarely, if every, travel off FOB.

period. Although there was some reduction in total training and preparation time for counterinsurgency and security force missions in 2008–2010 relative to 2003–2007, there was a bigger shift from post- to premobilization training in response to the limit on total mobilization

**Figure 3.1
Changes in Pre- and Postmobilization Training and Preparation Time for
BCTs**

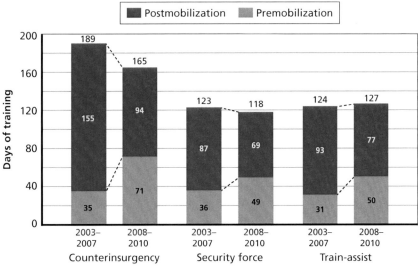

time. The results for enabler units, across deployment categories, shown
in Figure 3.2, are similar to those for BCTs.

We also examined predeployment training requirements, based
on FORSCOM guidance, First Army's Individual Training Tracker,
and data from First Army TSBs on how many soldiers had completed individual training requirements before mobilization. These
data come from 12 BCTs (including 101 individual units), 39 ARNG
enabler units, and 28 USAR enabler units (including engineer, military
police, quartermaster, and truck companies) that mobilized from 2008
through March 2010.

Units that deployed in support of OEF and OIF had to meet
extensive predeployment training requirements developed over time by
FORSCOM and the combatant command. The following list is based
on 2010 individual requirements:

Figure 3.2
Changes in Pre- and Postmobilization Training and Preparation Time for Enabler Units

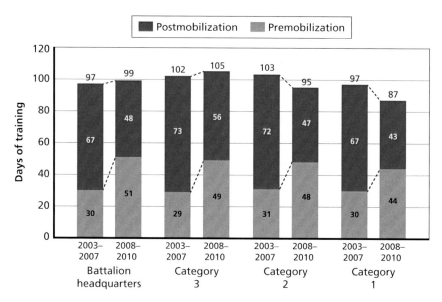

RAND *RR738-3.2*

- SRP (e.g., personnel and administration, finance, security clearances, wills, medical, dental, schooling)
- 30 briefings (including language and cultural familiarization, combat stress, suicide prevention)
- weapon qualification on one or more individual and crew-served weapons
- 32 Army Warrior Training tasks (categorized as shoot, move, communicate, survive, adapt)
- 12 battle drills (such as conducting patrols, reacting to contact, and evacuating casualties)
- 22 leader tasks (for NCOs and officers), plus five mission-dependent tasks.

Some BCTs scheduled training on collective tasks before mobilization, such as base defense, counter-IED procedures, checkpoint operations, and operations in urban terrain. However, most units planned to use the postmobilization period for the bulk of their collective training and for training focused on the unit's specific mission.

Table 3.2 shows an example of the data collected by First Army on premobilization individual training accomplishments for an ARNG military police company with a total of 185 authorized personnel. For each training task, the table shows the objective agreed at the joint assessment IPR, the number and percentage of soldiers who had completed all the tasks in that category as of the mobilization station arrival date (MSAD), and the number and percentage of soldiers who would have to complete the remaining requirements after mobilization. As Table 3.2 indicates, almost all the soldiers in this unit had completed their weapon qualifications when the unit reached the mobilization station, but substantial numbers had not completed the other individual requirements.

To summarize the premobilization individual training accomplishments for the sample of BCTs and enabler units that mobilized from 2008 to 2010, we calculated the percentage of all unit members

Table 3.2
Premobilization Individual Training Accomplishments for an ARNG Military Police Company

Task	JA Objective		MSAD Actual		Remaining	
	%	Number	%	Number	%	Number
Individual weapon qualification	100	257	99	256	1	1
Crew-served weapon qualification	100	43	100	43	0	0
Theater-specific individual training	100	185	64	118	36	67
Theater-specific leader training	100	75	73	55	27	20
Army warrior training	100	185	61	112	39	73
Battle drills	100	185	61	114	39	71
Combat lifesaver	100	185	84	156	16	29
Driver's training	75	138	81	112	19	26

who had completed all individual training requirements at the time the unit arrived at the mobilization station (Figure 3.3). In the median BCT unit, 71 percent of soldiers had completed all the required tasks. The median ARNG enabler unit had a similar result (70 percent), while the median USAR enabler unit did somewhat better (79 percent). Figure 3.3 also shows two other points on the distribution of units, the 25th and 75th percentiles. For example, this indicates that, in 25 percent of BCT units, 65 percent or less of the soldiers had completed all required tasks; in the top 25 percent of units, 80 percent or more of the soldiers had qualified. These results show considerable variability in the ability of RC units to complete individual premobilization training requirements.

Thus, despite the resources dedicated to increasing premobilization training after 2008, RC units found it difficult to complete all individual training requirements before mobilization. In most units, 20 to 30 percent of the soldiers still needed to complete some individual training after mobilization. Lippiatt and Polich, 2013, cites several factors that affected a unit's ability to complete required individual training tasks before mobilization:

Figure 3.3
Individual Training Accomplishment Before Mobilization

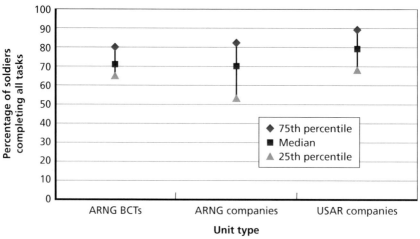

- **Equipment:** Soldiers must qualify on their weapons while wearing the most up-to-date body armor, which may not have been available to all units. Units may also have lacked night-vision equipment, counter-IED equipment, or the types of weapons required in theater or may not have had access to the kinds of vehicles used in theater, such as mine-resistant, ambush-protected vehicles (MRAPs).
- **AT attendance:** Some soldiers may have been unable to attend AT, when many of these tasks were trained. Historically, only 70 to 80 percent of unit members attend AT. Units may have also scheduled ATs longer than 15 days,[9] so some soldiers may not have attended the entire period due to civilian job commitments.
- **Personnel turnover:** Many units receive a large influx of new members in the last few months before mobilization. These newcomers may have missed some training events that were scheduled before they were assigned to the unit.

The USAR's RTCs may have mitigated some of these problems, enabling more USAR unit members to complete individual training tasks than in comparable ARNG units (see Figure 3.3). The RTCs had the same types of equipment used in theater, necessary ranges and training areas with training lanes set up, and observer controller-trainers to facilitate field events, such as weapon qualification, Army Warrior Training, and battle drills. To reduce problems associated with personnel turnover, RTC rotations were typically scheduled close to the unit's mobilization date. These rotations included 17 training days out of a total of 21 days.

However, it should be noted that, due to the close coordination between RC units and First Army established under EXORD 150-08, both ARNG and USAR units successfully made up training shortfalls after mobilization, without postponing their latest arrival date in theater.

---

[9] Lippiatt and Polich, 2013, p. 38, reports that soldiers in the median unit in their sample put in an average of 34 to 36 AT days in the year before mobilization, in comparison to a norm of 15 days for units not preparing to deploy. Typically, this was done in two separate AT periods, one lasting about two weeks and another lasting about three weeks.

Unit Evaluations of First Army Training Support

This section summarizes our review of a small sample of the AARs RC units submitted to First Army approximately 90 days after deployment into theater. First Army asked units to send these AARs to provide feedback on the training process and the quality of the training that was provided, how well the training was executed, and whether it appropriately prepared soldiers for their mission in theater. The AARs we received and analyzed provide some deeper (although less quantifiable) insights into the challenges units faced before, during, and immediately after their mobilizations.

First Army sent us a small sample of nine AARs written in 2009 and 2010 that were drawn from a mix of brigade-sized combat arms units and support units. These units deployed to OEF, OIF, and Kosovo Force, an international peacekeeping mission. These reviews were in PowerPoint format and addressed issues that unit leaders felt would be useful for First Army to consider for improving its efforts to support units during the mobilization process, but there was no standardized format for the AARs. In addition, we reviewed a CALL newsletter that focused extensively on the mobilization and deployment of an ARNG infantry BCT (CALL, 2009).

We gleaned three general findings from our review of these sources:

- Most RC unit concerns were similar to comments AC units preparing for deployment made.
- Some concerns expressed in the AARs were linked to reserve status. In particular, personnel databases and medical and dental readiness issues were exacerbated by reserve status.
- AARs are not the best tool for evaluating predeployment training.

We discuss each of these findings in more detail below. Although it must be noted that this is a small sample of dissimilar products (the quality of the AARs will be discussed later in this section), they provide useful context for considering the overall conclusions of this study.[10]

Most RC Comments Were Similar to AC Concerns

Most of the issues addressed in the AARs were related to the quality of training and its relevance to the unit's mission in theater (where these AARs were written). One common AAR topic was the dearth of theater-specific equipment for use in pre- and postmobilization training, such as command, control, intelligence, surveillance, and reconnaissance systems; databases; and MRAPs. Some units commented that pre- and postmobilization training did not adequately address the most current tactics, techniques, and procedures.

Other units recommended improving synchronization and coordination between incoming units and the ones they were relieving in theater. Some AARs suggested that First Army could facilitate video teleconferences and email introductions to get the most up-to-date information on the local battlespace, as well as best practices and updated tactics, techniques, and procedures. Personnel at First Army told us that these concerns have since been addressed to some degree. For example, First Army has liaison officers located in theater to provide feedback on current operations and recommend adjustments to predeployment training.

These concerns related to the quality of training are similar to those AC units raised about their own predeployment training (CALL, 2008). Thus, from the trained unit's perspective, the most significant issues regarding predeployment training were not directly related to how well First Army supported the training process.

Issues Related to RC Status

Although many concerns highlighted in AARs were common to both AC and RC units, some issues were specific to RC units. One concern was that logistics and administrative tasks related to the mobilization process interfered with training. One unit (an ad hoc adviser

---

[10] We will not refer to specific units in our analysis below, but our list of references includes the sample of AARs.

unit) noted that the deployment workload for its logistics (S-4) section in managing unit movement, operational needs statements, and additional tasks was necessary but distracted soldiers from their postmobilization training. This unit suggested that First Army should help facilitate some of these logistics tasks so that soldiers can concentrate on postmobilization training.

Another concern was that premobilization medical and dental readiness was difficult to achieve. In particular, some RC units are unable to access military treatment facilities to bring all personnel up to mandated medical and dental readiness levels. Relying on civilian clinics required much negotiation and coordination with civilian care providers. Despite significant flexibility on the part of civilian providers, current policies, confusion about standards, and other coordination issues resulted in the unit having to devote significant time and resources to ensure that the unit mobilized with as many qualified personnel as possible (CALL, 2009).

Another friction point noted was that units were unable to access and properly use some AC databases (e.g., the Army Training Requirements and Resources System and Global Interactive Personnel Electronic Records Management System) necessary for processing newly joined personnel, including soldiers being mobilized from Individual Ready Reserve status who required MOS retraining (hence, the need to access the Army Training Requirements and Resources System). In particular, the unit commented that there was no clear guidance on how to use these databases and applications when dealing with RC personnel. This concern highlights a possible gap in the Army's ability to integrate RC units and personnel into its administrative systems. Notably, this problem had also occurred roughly 20 years earlier, when RC units were mobilized to support ODS and faced similar problems accessing and integrating data into AC automated systems.

Finally, some units noted that training events were needlessly repeated as the unit moved through the mobilization process. In particular, one unit had completed certain training events at home station, which were then repeated at the RTC, and again at the mobilization station. Unit members observed that, although they had already completed the training event, a general officer–level waiver was required for them to avoid repeating the training event at the RTC (and subsequently, at the mobilization site). However, given the amount of unit personnel turnover that likely occurred in these units in the months before deployment, some soldiers might have needed to perform these training tasks at the RTC or the mobilization station.

It is also important to note that these AARs were written in 2009 and 2010. First Army personnel indicated that they have since addressed or mitigated some of these issues. Repetitive training is an important concern, although the AC must continue to validate RC training to maintain consistency. However, the nature of the issues raised does not seem to require extensive changes in legislation or in organizations that support RC training and mobilization.

Using AARs as an Evaluation Tool

The third observation from our review of First Army AARs is that the AAR process itself is not a well-developed tool for evaluating postmobilization training. Army doctrine on AARs focuses on their use in providing feedback on a unit's performance during evaluated field exercises (i.e., exercises using observer controllers) (U.S. Army Combined Arms Training Center–Training, 2011). AARs are being used in many more applications and situations than field exercises, but this proliferation has not been accompanied by a similar proliferation of AAR methods, practices, or metrics.

Specifically, the AAR is a qualitative tool that has very little formal framework. This is a good feature in field exercises, allowing the unit to quickly capture thought processes immediately after an exercise to understand why certain decisions and actions unfolded as they did (Salter, 2007). It has also been used successfully in other contexts. However, it is primarily intended to evaluate the unit's performance during a collective training event and is less useful for evaluating the content of postmobilization training, which has specific associated metrics and goals, or the training support First Army provides. In addition, the AAR's inherent qualitative nature results in evaluations that are not easily comparable. This is not a problem when AARs are used in their original role, but it makes them less useful for examining trends across units and over time.

This was certainly the case in the sample of AARs provided by First Army. Units were given very little guidance on how to structure their AARs, other than to focus on issues that they felt were important. While the result has allowed First Army to make incremental refinements to its practices, the inability to aggregate AARs and make direct comparisons across units means that AARs cannot be easily used to make generalizations or to add depth and context to the quantitative metrics we used to measure pre- and postmobilization training.

Implications of Army Plans for Future RC Training

Because U.S. deployments to Iraq and Afghanistan have declined, the Army must make decisions about future RC training requirements and training support. Due to declining AC end strength and budget constraints, the Army will likely need to continue to rely on the ARNG and USAR as an operational reserve to be able to expand capabilities rapidly if the demand for forces suddenly rises. The international security situation remains complex, so it is difficult to predict when and where the next global contingency will occur and, thus, what unit types and theater-specific predeployment training will be needed.

This chapter describes the Army's plans for RC force generation and training aim points as of FY 2013, when we conducted interviews and gathered supporting documents from the Army organizations involved in AC support for RC training, including the Office of the Deputy Chief of Staff, G-3/5/7; FORSCOM; USARC; NGB; and First Army. The Army's plans were evolving during the course of our study and are likely to continue to evolve in response to changes in force structure and resources. However, we have drawn some broad implications for the future structure of RC training support from our historical analysis and these plans.

Future RC Training Requirements

As of FY 2013, the Army planned to continue its ARFORGEN cyclical readiness process for both AC and RC forces, with AC units on a 24-month cycle and RC units on a 60-month cycle. Figure 4.1 shows

Figure 4.1
RC Force Generation Process

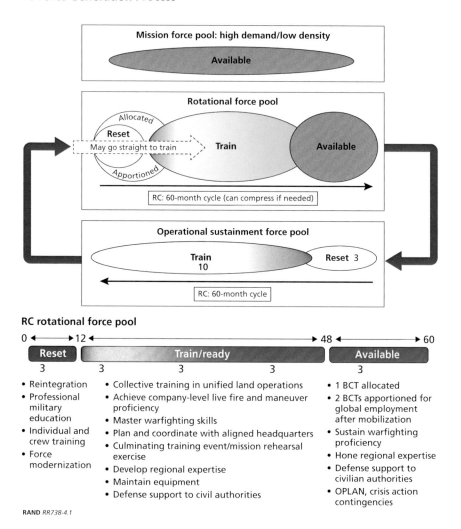

RAND *RR738-4.1*

the force generation process for RC units, based on U.S. Army, 2012. The Army plans to divide RC units into three force pools:

- The Mission Force Pool consists of high-demand, low-density unit types, when the total number of units is too small to support rotations or when the units are assigned to specific theaters.

These units must continuously sustain a high readiness rating, as specified by the supported combatant command. Examples of AC units in this category are units assigned to Korea and theater-assigned unique capabilities (referred to as Theater Committed, Nonrotational). An example of an RC unit in this category is the 189th Transportation Company, which is expected to be always available to support U.S. Northern Command.

- The Rotational Force Pool consists of units needed to meet known rotational requirements or to execute the first rotation in support of a designated OPLAN. These units will follow the ARFORGEN cyclical readiness process. When these units reach their available year, they can be deployed in support of named operations, non-programmed combatant command requirements, or theater operational exercises or training. Units that are not deployed during their available year are distributed to combatant commands as regionally aligned forces based on specific OPLANs.

- The Operational Sustainment Force Pool consists of RC units that will not be needed to deploy until the second or third rotation of an operation, or Phase IV of an OPLAN. These units will be in a cyclical readiness process, but will not be resourced to attain as high a level of readiness as those in the Rotational Force Pool in their available year. It is expected that there will be sufficient time after a contingency event occurs to prepare these units for a later rotation (FORSCOM, 2013).[1]

The bottom panel of Figure 4.1 shows planned activities for BCTs in the Rotational Force Pool during each phase of the ARFORGEN process.[2] These plans suggest that BCTs would focus primarily on collective training activities during the train/ready phase and achieve company-level live fire and maneuver proficiency by the

---

[1] These units can also provide defense support to civil authority missions, or transition into the rotational pool, if needed, to meet overseas deployment requirements or to modernize equipment as part of Army equipment modernization programs (U.S. Army, 2012).

[2] Note that the figure indicates there would be three BCTs in each year of the Rotational Force Pool and 13 BCTs in the Operational Sustainment Force Pool—ten in the train phase and three in the reset phase.

time they reached their available year. However, past experience with Bold Shift and RC units preparing to deploy to support OEF and OIF indicates that RC units struggled to complete individual and crew- and squad-level training requirements during the premobilization period, even with additional resources dedicated to training and training support. RC units face many difficulties in meeting premobilization training requirements. Many of the same problems observed in the ARNG roundout brigades in 1990 and 1991 persisted in units preparing to deploy for OEF and OIF more than ten years later, although some were less severe. These problems included the limited number of premobilization or peacetime training days, access to training ranges and maneuver areas, availability of up-to-date equipment, low AT attendance, and high personnel turnover. For the most part, these are longstanding problems that are expensive and/or difficult to resolve.

Furthermore, if SRP and other individual training requirements are not met prior to mobilization, these tasks will have to be completed after mobilization, and would affect the types of postmobilization training support that will be needed. The Army is currently revising AR 350-1, Army Training and Leader Development, which was last published in December 2009.[3] The new regulation will specify Army-wide training requirements, which may exclude some theater-specific training requirements that were needed for OEF/OIF. However, many of the training requirements for recent operations are annual requirements that units should conduct whether deploying or not, such as Army Warrior Tasks and Battle Drills (15 individual tasks and six battle drills), mandatory annual briefings, and mission-essential tasks (FORSCOM, 2013). In addition, U.S. Army G-3/5/7, 2012, indicates that the number and length of common mandatory training tasks has increased in such areas as resilience training, substance abuse, sexual harassment and assault response and prevention, Army values, and information security.

---

[3]   A rapid action revision of AR 350-1 was issued in August 2011 to implement the Don't Ask, Don't Tell Repeal Act of 2010 and make some administrative changes, such as correcting forms, publication titles, and website addresses (AR 350-1, 2011).

Thus, the assumption that RC combat units will be able to achieve company-level proficiency by the end of the train/ready phase may be overly optimistic, and if units focus on collective training, many SRP and individual training requirements may still need to be completed after mobilization.

Plans for RC Training Support

During the interviews conducted for this study, we collected information from FORSCOM and First Army on their plans for future RC training support. Table 4.1 shows FORSCOM's plans for training and readiness oversight across the ARFORGEN cycle, primarily for Rotational Force Pool units (FORSCOM, undated). This time line assumes that units scheduled to deploy during their available year would receive a notification of sourcing two years before mobilization (at the beginning of Train/Ready 2). Even for units that are not planned to deploy, training and readiness oversight during Train/Ready 3 become particularly important as part of the validation process, to reduce post-mobilization training time for RC units required early in an OPLAN.

Based on the Army's RC force generation plans, First Army projects that it will need to support training events for 70,000 RC soldiers in the Rotational Force Pool per year. To fully support this throughput of soldiers (including all deploying units and 32 training events for distributed, nondeploying units), First Army estimates that it would need 3,075 trainer/mentors. At lower levels of capability, 2,568 trainer-mentors could support all deploying units and 27 FORSCOM priority 1 and 2 training events for nondeploying units,[4] and 1,968 trainer-mentors could support all deploying units and 17 FORSCOM priority 1 training events for nondeploying units (First Army, 2013b).

---

[4] FORSCOM priorities are defined as follows: (1) BCTs (regionally aligned forces or aligned with a combatant commander's OPLAN), (2) USAR Level 1 and 2 units aligned with a combatant commander's OPLAN, (3) ARNG combat aviation brigades, (4) ARNG fires brigades, (5) RC functional and multifunctional brigades (USAR and ARNG), and (6) division headquarters (First Army, 2013b).

Table 4.1
Training Support and Deployment Validation

Unit Type	Reset	Train/Ready		
		1	2	3	Available
Rootational • Allocated • Apportioned/ distributed	Advise or assist on training plan, review USR	Observe yearly training brief, review USR	Joint Assessment Conference	Support or observe collective training events	Validate deployable or available
	Provide training support as requested by unit during training synchronization conference				
Operational sustainment	Advise or assist training as requested by unit (within capability)				

SOURCE: FORSCOM, undated.

NOTE: First Army synchronizes RC pre- and postmobilization training exercises through a series of conferences, including the First Army Training Support Synchronization Work Group, FORSCOM's ARFORGEN Synchronization and Resourcing Conference, and Joint Assessment Conferences (for mobilizing units) (First Army, 2014).

Figure 4.2 shows the total number of Title XI personnel authorized and assigned, as well as the portion authorized and assigned to FORSCOM or First Army, based on Army posture statements from 2004 through 2012.[5] As Figure 4.2 indicates, the number of authorized Title XI positions tracks closely with the legal requirement, which declined from 5,000 to 3,500 starting in 2005. This change was implemented by the Army over a three-year period. However, in the years for which data were available, only about 84 percent of these authorized positions were filled (on average), and First Army expects its fill rate to fall to 80 percent by the end of FY 2014. Of the 3,299 Title XI positions authorized at First Army, 2,460 are trainer-mentor positions. (The remaining 839 are mission command and support positions.) At an 80-percent fill rate, there will be 1,968 Title XI trainer-

Figure 4.2
AC Personnel Authorized and Assigned Under Title XI

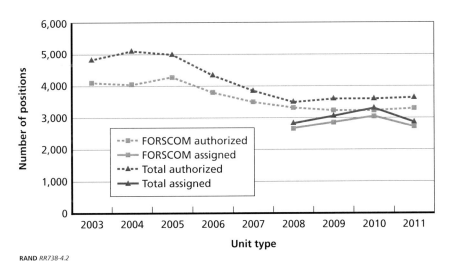

RAND RR738-4.2

---

[5] As of 2011, approximately 650 Title XI positions were authorized in the USAR, TRADOC, and U.S. Army Pacific Command (McHugh and Odierno, 2012).

mentors, enough to support deploying RC units and FORSCOM priority 1 training events.[6]

In addition to Title XI AC personnel, First Army also has AGR, military technician, and Department of the Army civilian personnel, as well as USAR troop program units, including TSBns and logistics support battalions (LSBns), under its operational control. Table 4.2 provides a snapshot of authorized and assigned personnel as of February 2013. During recent operations, RC soldiers on Contingency Active Duty for Operational Support (Co-ADOS) orders and temporary civilian hires have been used to support mobilizations.[7] The number of Co-ADOS positions peaked at about 6,800 in 2009 but has been declining in recent years, and all temporary positions are expected to be eliminated by the end of FY 2014. RC TSBns and LSBns can provide peacetime training support during their AT periods and expansion capacity to meet OPLAN surge requirements. Other USARC organizations providing training support include the 84th Training Command, which provides lanes training, and the 75th Mission Command Training Division, which provides command and staff training.

Note that, as of February 2013, the ARNG had filled less than a third of its 215 authorized AGR positions in First Army. USAR TSBns and LSBns (during their AT periods) could also be a source of the trainer-mentors needed to augment the number of Title XI trainer-mentors available to support RC training events for nondeploying units.

In FY 2014, First Army began to implement "Operation Bold Shift" to support its change in focus from postmobilization training back to premobilization training and to streamline training support organizations to reduce costs. Bold Shift will reorganize the TSB structure to form fewer, but more-capable TSBs, increase the number of trainer-mentors by 32 percent, and adjust the MOS mix to increase

---

[6] An additional 600 trainer-mentors would be needed to support priority 2 training events. A total of 1,107 additional trainer-mentors would be needed to support all 32 training events for nondeploying units.

[7] Under the Operation Warrior Trainer program, USAR and ARNG soldiers who had recently returned from a deployment could volunteer to remain on active duty for one to three years to train mobilizing soldiers (Witscheber, 2006).

Table 4.2
First Army Positions as of February 2013

Type of Personnel	Authorized	Assigned
AC (Title XI)	3,299	3,075
USAR AGRs	409	399
ARNG AGRs	215	61
TPUs (TSBns, LSBns)	7,075	6,113
Military technicians	84	62
DA civilians	386	349
RC on Co-ADOS orders		1,383
Temporary civilians		130

SOURCE: First Army, 2013b.

capability to support collective training for both combined arms maneuver and functional or multifunctional forces. USAR TSBns and LSBns will be restructured to return redundant capabilities back to the USAR. TSBs will also have the capability to provide simulation-based mission command staff training and design exercises at battalion and higher levels (First Army, 2014).

Objectives for Future RC Training Support

Our review and analysis of past experience and interviews with Army commands and other organization involved with RC training support suggests four broad objectives that future training support arrangements should meet:

1. **All components should train to well-defined, common standards for similar unit types and missions.** Army-wide requirements will be defined by the revised AR 350-1, while combatant commands may add theater-specific requirements for deployed or regionally aligned forces. This objective also supports Army Directive 2012-08, Army Total Force Policy (McHugh, 2012), which requires Army commands and Army service component

commands to ensure that the procedures and processes for validating the predeployment readiness of assigned forces are uniform for AC and RC units and soldiers.

2. **Establish well-defined, integrated pre- and postmobilization training requirements for force pools within the ARFORGEN process.** Training requirements should be informed by historical data on required training times, AT attendance, and personnel turnover and take into account constraints on RC training time and resources. Integration of premobilization and postmobilization training requirements and oversight is needed to shorten postmobilization training and validation times.

3. **RC training capability must be maintained in peacetime, and capacity must be rapidly expandable to meet surge requirements.** Capability should be exercised in peacetime, including First Army and USAR TSBns and LSBns.[8] Peacetime capacity should be driven by the expected peacetime throughput of units. Capacity must also expand to meet OPLAN demands for throughput of RC units by type. Support organizations should have high-quality trainer-mentors and the right mix of MOSs to support expected throughput of units. Training infrastructure must also be expandable, since mobilization stations (such as Camp Shelby) may be minimally staffed and equipped during peacetime.[9]

4. **Establish mechanisms to evaluate and improve training support.** First Army should expand and improve its current evaluation process to obtain better feedback from supported RC units and from receiving combatant commands, when appropriate.

---

[8] The Army could also consider exercising its surge capacity every few years.

[9] As RC mobilization requirements have declined, First Army has reduced the number of mobilization force generation installations to three (Fort Hood, Fort Bliss, and Joint Base McGuire-Dix-Lakehurst). It maintains a footprint at four inactive sites (Camp Shelby, Camp Atterbury, Joint Base Lewis McChord, and Fort Stewart) to respond to mobilization surge requirements. It routinely reviews the inactive sites to identify mobilization training and capacity requirements, shortfalls, and ability to support contingency mobilization surge requirements (First Army, 2014).

CHAPTER FIVE

Conclusions and Recommendations

This chapter summarizes the conclusions of the study and our recommendations. We also review the major provisions of Title XI and discuss which sections may need to be revised to reflect changes in RC training requirements and training support since the early 1990s.

RC Training Requirements and Training Support

Congress established Title XI to address specific problems observed during the mobilization of three ARNG roundout brigades to support ODS. These problems were documented in reports from the Army Inspector General, GAO, Congressional Research Service, and other sources. Most of the individual provisions of Title XI can be traced directly to concerns expressed in these reports and in congressional testimony associated with Title XI. For example, Congress wanted to increase the quantity and quality of full-time support personnel available to assist RC units with training; improve the accuracy of readiness ratings; and focus premobilization training at the individual, crew, and squad levels.

Many of the same problems observed in the roundout brigades in 1990 and 1991 persisted in RC units preparing to deploy in support of OEF and OIF, although some were less severe. These problems included a limited number of training days available during the premobilization period; limited access to training ranges, maneuver areas, and some types of equipment (such as up-to-date body armor, night-vision goggles, and MRAPs); and personnel turnover and AT attendance issues that limit the effectiveness of premobilization training. RC units typically required 45 to 50 training days just to meet individual preparation and training requirements for deployment.

Thus, the historical evidence suggests that premobilization training should focus on individual soldier qualifications and collective training at the crew, squad, and platoon levels, particularly for combat units, and on mission command training for battalion and higher staffs. Some company-level training may be feasible in Train/Ready Years 2–3 of the ARFORGEN cycle for enabler units and as time permits for combat units.

The Army currently has a multicomponent RC training support structure that has worked relatively well in support of OEF and OIF. The "peacetime" training support structures that were in place in 2001 were adjusted over time to handle the increased volume of RC mobilizations. First Army's resources were diverted to postmobilization training support and augmented by mobilized reservists and temporary civilian hires. After 2008, when the Secretary of Defense limited RC mobilizations to one year, the ARNG and USAR increased premobilization training support, establishing PTAEs and RTCs, primarily using overseas contingency operations funding. As the number of RC mobilizations is declining, First Army is reestablishing premobilization training support relationships with RC units and providing support in Train/Ready Years 2–3, which is needed for training and readiness oversight and to shorten postmobilization training and validation time lines.

A unified, multicomponent training support organization is consistent with DoD and Army Total Force policies and reduces the chance that training standards will diverge across components in the future. However, First Army may need to make greater use of USAR training support personnel during their AT periods to support premobilization collective training events. It also appears that the ARNG is relatively underrepresented in First Army and could help increase its involvement by filling its existing AGR positions. In addition, First Army may need to review the MOS mix of its trainer-mentors to ensure that it reflects the expected peacetime throughput of RC units. First Army has indicated that it is implementing an ARFORGEN cycle for USAR TSBns, in coordination with USARC, to provide unit-based training support. It is also adjusting the MOS mix of trainer-mentors and recommends a collaborative periodic review by First Army, NGB, and USAR to adjust authorization and manning levels to ensure effective training support (First Army, 2014).[1]

First Army established an AAR process to obtain feedback from RC units on predeployment training and areas for improvement. This process could be significantly improved to provide better feedback during peacetime, as well as during future large-scale mobilizations. Some actions could be taken immediately within existing resources to improve the current evaluation process, such as the following:

- **Make feedback comparable across units,** possibly using a two-part format. The first part would consist of a standard list of questions to collect data that are comparable across units and over time that could be implemented in an automated, web-based format. The second part would allow more of a "free response" format similar to the current process to capture issues that are most salient to units and of which First Army may not be aware. First Army might also consider seeking feedback from the higher headquarters receiving the RC unit during deployment, although this information might be considered sensitive.
- **Spread out the feedback process over time** by administering one review at the end of postmobilization training and another 70 to 90 days after deployment. This would reduce the reporting burden on the units after deployment and yield more-detailed results about the mobilization process. It would also allow First Army more time to implement any changes to the mobilization process based on the comments it receives.
- **Make the results of the feedback process easily accessible** to both First Army planners and units going through the mobiliza-

---

[1] First Army also noted in its comments on an earlier draft of this report that, regardless of which Army component or organization provides premobilization training support, adequate resources must be provided to ensure that mandated training requirements are met, that Army standards are adhered to, and that RC forces are being trained and validated to accomplish assigned missions in support of combatant commanders.

tion process to help improve the quality of training from both the provider and unit perspectives. This goal could be facilitated by using a web-based format. Another possible approach might be to submit the data collected by First Army to CALL to manage and analyze historical trends.

- **Consider "prebriefing" units and training support providers before mobilization** to orient them with the problems and obstacles that previous units of similar types have faced during the mobilization process.

Over the longer term, First Army could establish a more-deliberate and concurrent evaluation and improvement process by creating a process improvement cell in First Army G-3 (Future Operations) whose primary duty would be to collect and manage the feedback process and advise the First Army G-3 on changes and adjustments to the mobilization process to improve training support and reduce delays in postmobilization training and validation.

Changes to Title XI

Title XI is now over 20 years old, and some of its provisions no longer reflect the current operating environment. Others remain relevant. As for the remainder, it is unclear whether they should be retained or not, depending on the expense or difficulty of compliance and how the Army decides to structure its future RC training support organizations. Table 5.1 summarizes the major provisions of Title XI and some considerations for change.

While it is still important for the AC to be involved in RC training, it is not clear what the "right" number of AC personnel to be assigned to First Army and other RC training support organizations should be. First Army has estimated the number of trainer-mentors needed to support RC collective training exercises, but this number could change as force structures and budgets change, and they do not all need to be AC personnel. A more-flexible approach might be to specify the proportion of AC personnel assigned as trainer-mentors or

Table 5.1
Considerations for Changes to Title XI

Title XI Provisions (as amended)	Considerations for Change
§1132—Requires 3,500 AC personnel to be assigned as advisers to RC units	Maintains AC involvement in RC training Not clear what the "right" number is Consider adding ARNG and USAR personnel
§1131—Requires all ARNG combat units and high-priority USAR and ARNG CS and CSS units to be associated with an AC unit	No longer relevant First Army executes roles and requirements assigned to commanders of associate units
§1119—Establish a program to minimize postmobilization training time. Pre-mobilization training should focus on individual, crew/squad, and platoon levels.	Still relevant, based on historical evidence and recent RC mobilizations
§1135—Identify priorities for mobilization of RC units and specify the required number of postmobilization training days	Still relevant, may need to be modified to reflect ARFORGEN (cyclical versus tiered readiness)
§1121—Modify the RC readiness rating system to provide a more accurate assessment of deployability and resource shortfalls	Army has modified readiness ratings, but measuring readiness objectively is still problematic
§1120—Expand use of simulations, simulators, and advanced training devices to increase training opportunities for RC units	Still relevant
§1111—Increase percentage of RC personnel with prior AC experience to 65% for officers and 50% for enlisted (Modified in 1996 to numerical goals for transition of 150 officers and 1,000 enlisted personnel annually from AC to National Guard through specific programs)	Although AC experience is still valuable, the original goals would have been difficult and potentially costly to reach Modified provision is no longer relevant, because programs mentioned in legislation may no longer exist Many current RC personnel have deployment experience, but it will recede over time
§1114—Military education requirements for NCOs must be met for promotion to a higher grade	Still relevant

Table 5.1—Continued

Title XI Provisions (as amended)	Considerations for Change
§1115—Establish a personnel accounting category for ARNG members who are not available for deployment	Army tracks number who are nondeployable, but ARNG and USAR have small TTHS accounts that do not include all nondeployable personnel
§1134—Requires report on compatibility of AC and RC equipment and effects on combat effectiveness	Still relevant
§1134—Ensure that personnel, supply, maintenance management, and finance systems are compatible across all components	Still relevant

elsewhere to support RC training so that the number can be adjusted with changes in force structure or training demand. It might also be appropriate to specify the proportions of ARNG and USAR personnel assigned to these roles.

First Army and other multicomponent training support units have been used to fulfill Title XI's requirements for associate units. As the 2012 Army Posture Statement notes, the "formal training relationships previously established by the AC/RC Association Program … were modified as the requirements of ongoing Overseas Contingency Operations kept AC units in frequent deployments and RC units in frequent mobilization." Instead, the congressional intent was met using AC-led multicomponent units (such as First Army) to provide the necessary contact with mobilizing RC units and execute the other roles and responsibilities formerly given to the commanders of associate AC units (McHugh and Odierno, 2012).

Two other provisions listed in Table 5.1 are also candidates for change. First, the Army had difficulty meeting the goal of increasing the percentages of RC officers and enlisted personnel with AC experience.[2] This provision was modified by the NDAA for 1996 to reflect

---

[2] Based on Army Posture Statements from 2002 through 2012, neither the ARNG nor the USAR has met these goals more recently. For example, in 2011, 30 percent of enlisted personnel and 49 percent of officers in the ARNG had AC experience; 21 percent of enlisted personnel and 33 percent of officers in the USAR had AC experience.

numerical goals of 150 officers and 1,000 enlisted personnel with AC experience joining the ARNG each year, but refers to specific transition programs that may no longer exist.[3] In any case, many current RC personnel now have deployment experience, although this experience will recede over time as the number of mobilizations decreases.

Second, the Army tracks the numbers of ARNG and USAR personnel who have not completed initial entry training requirements or are otherwise nondeployable. However, the ARNG and USAR only have small TTHS accounts that do not include all nondeployable personnel. To increase the size of their TTHS accounts, ARNG and USAR would either have to give up some other force structure to remain within end strength limits or have to fund the additional positions in the TTHS accounts. Maintaining an RC TTHS account is also administratively more difficult than in the AC, particularly for the ARNG, which is organized at state level, because soldiers are recruited to a unit in a specific geographic location and expect to return to the same unit when they complete training or when their deployment-limiting condition is resolved.[4]

Most of the other Title XI provisions listed in Table 5.1 remain relevant, but may require some revision, for example, to reflect the ARFORGEN concept of cyclical, rather than tiered, readiness. The remaining provisions are discussed in the appendix.

---

[3] Section 514 of the NDAA for 1996 refers to a program that permits the separation of officers on active duty with at least two, but fewer than three, years of active service if the officer is accepted for appointment in the ARNG. The provision for enlisted personnel refers to a program established by the Secretary of Defense in 1991 to test the cost-effective use of special recruiting incentives involving not more than 19 noncombat arms skills approved in advance by the secretary.

[4] An alternative proposed by the Army Inspector General's report was to increase personnel fill up to 110 percent of authorized strength. While such a policy might be targeted at RC units in Train/Ready year 3 of the ARFORGEN cycle, it could be difficult to achieve if it required moving soldiers across states or geographical regions. Moreover, Lippiatt and Polich (2010, p. 38) found that ARNG infantry battalions that deployed between 2003 and 2008 were manned at an average of about 125 percent of authorized strength at their mobilization date, but deployed at about 85 percent of authorized strength. Increasing personnel fill above authorized strength would also require the ARNG and USAR to either give up force structure or fund the additional positions.

Title XI and Related Legislation

This appendix summarizes the provisions of Title XI of the NDAA for 1993 and other related legislation, as well as related sections of U.S. Code.[1]

National Defense Authorization Act for 1992 and 1993 (PL 102-190)

Section 414: Pilot Program for Active Component Support of the Reserves

(a) During fiscal year 1993, the Secretary of the Army shall institute a pilot program to provide AC advisers to combat units, combat support units, and combat service support units in the Selected Reserve that have a high priority for deployment. The advisers shall be assigned to full-time duty in connection with organizing, administering, recruiting, instructing, or training such units.

(b) The objectives of the program are as follows:

    (1) To improve the readiness of units in the Army reserve components

    (2) To increase substantially the number of AC personnel directly advising RC unit personnel

---

[1] Note that we have paraphrased or omitted some text, so this is not a verbatim copy of the legislation.

    (3) To provide a basis for determining the most effective mix of RC and AC personnel in organizing, administering, recruiting, instructing, or training RC units

    (4) To provide a basis for determining the most effective mix of AC officers and enlisted personnel in advising RC units regarding organizing, administering, recruiting, instructing, or training RC units.

(c) Personnel to be assigned:

    (1) The Secretary shall assign officers, warrant officers, and enlisted members to serve as advisers under the program.

    (2) The Secretary shall assign at least 1,300 officers as advisers to combat units and 700 officers as advisers to CS and CSS units.

(d) Based on the experience under the pilot program, the Secretary of the Army may expand or modify the program as he considers appropriate in order to increase the readiness and training of RC units for any period after September 30, 1993. Modifications in the program may not reduce the minimum number of officer advisers assigned below 2,000.

National Defense Authorization Act for 1993 (PL 102-484)

Title XI, Army National Guard Combat Readiness Reform Act of 1992
Subtitle A—Deployability Enhancements
Section 1111: Minimum Percentage of Prior Active Duty Personnel

(a) The Secretary of the Army shall have an objective of increasing the percentage of qualified prior active-duty personnel in the Army National Guard to 65 percent for officers and 50 percent for enlisted members, by September 30, 1997.

(b) The Secretary shall prescribe regulations establishing accession percentages for officers and enlisted members for fiscal years 1993 through 1997 so as to facilitate compliance with the objectives in subsection (a).

(c) Qualified prior active-duty personnel are members of the Army National Guard with not less than two years of active duty.

Section 1112: Service in Selected Reserve in Lieu of Active Duty Service

   (a) Academy Graduates and Distinguished Reserve Officer Training Corps (ROTC) Graduates

      (1) An officer who is a graduate of one of the service academies or who was commissioned as a distinguished ROTC graduate and is released from active duty before completing his active duty service obligation shall serve the remaining period of that obligation as a member of the Selected Reserve.

      (2) The Secretary concerned may waive this requirement if there is no unit position available for the officer.

   (b) ROTC Graduates. The Secretary of the Army shall provide a program under which graduates of the ROTC program may perform their minimum period of obligated service by a combination of two years of active duty and the remainder of their service obligation in the National Guard.

Section 1113: Review of Officer Promotions by Commander of Associated Active Duty Unit

   (a) Whenever an officer in an Army National Guard unit is recommended for a unit vacancy promotion to a grade above first lieutenant, the recommended promotion shall be reviewed by the commander of the active-duty unit associated with the National Guard unit, or another active-duty officer designated by the Secretary of the Army. The commander or other designated active-duty officer shall provide to the promoting authority, before the promotion is made, a recommendation of concurrence or nonconcurrence in the promotion, within 60 days after receipt of notice of the recommended promotion.

   (b) Subsection (a) shall take effect—

      (1) On April 1, 1993, for officers in ARNG units that are designated as round-out/round-up units;

      (2) On October 1, 1993, for officers in other ARNG units that are designated as early deploying units; and

(3) On April 1, 1994, for officers in all other ARNG combat units

Section 1114: Noncommissioned Officer Education Requirements

(a) Any standard prescribed by the Secretary of the Army establishing a military education requirement for NCOs that must be met as a requirement for promotion to a higher grade may be waived only if the Secretary determines that the waiver is necessary in order to preserve unit leadership continuity under combat conditions.

(b) The Secretary shall ensure that there are sufficient training positions available to enable compliance with subsection (a).

Section 1115: Initial Entry Training and Nondeployable Personnel Account

(a) The Secretary of the Army shall establish a personnel accounting category for ARNG members who have not completed the minimum training required for deployment or who are otherwise not available for deployment. The account shall be used for the reporting of personnel readiness and may not be used as a factor in establishing the level of Army Guard and Reserve force structure.

(b) Until an ARNG member has completed the minimum training necessary for deployment, the member may not be assigned to fill a position in an ARNG unit, but shall be carried in the account established under subsection (a).

(c) If at the end of 24 months after an ARNG member enters the National Guard, the member has not completed the minimum training required for deployment, the member shall be discharged. The Secretary may waive this requirement in the case of health care providers and in other cases determined necessary. The authority to make such a waiver may not be delegated.

Section 1116: Minimum Physical Deployability Standards

The Secretary of the Army shall transfer the personnel classification of an ARNG member from the member's unit to the personnel account established in section 1115 if the member does not meet minimum physical profile standards required for deployment. Any such transfer shall be made not later than 90 days after the determination that the member does not meet such standards.

Section 1117: Medical Assessments

The Secretary of the Army shall require that:

    (1)   Each ARNG member undergo a medical and dental screening on an annual basis; and

    (2)   Each ARNG member over the age of 40 undergo a full physical examination not less often than every two years.

Section 1118: Dental Readiness of Members of Early Deploying Units

  (a) The Secretary of the Army shall develop a plan to ensure that ARNG units scheduled for early deployment in the event of a mobilization are dentally ready for deployment.

  (b) The Secretary shall submit to the House and Senate Armed Services Committees a report on such plan not later than February 15, 1993. The report shall include any legislative proposals that the Secretary considers necessary to implement the plan.

Section 1119: Combat Unit Training

The Secretary of the Army shall establish a program to minimize the post-mobilization training time required for ARNG combat units. The program shall require:

  (1) That unit pre-mobilization training emphasize:

    a. Individual soldier qualification and training;

    b. Collective training and qualification at the crew, section, team, and squad level; and

    c. Maneuver training at the platoon level as required of all Army units; and

  (2) That combat training for command and staff leadership include annual multi-echelon training to develop battalion, brigade, and division level skills, as appropriate.

Section 1120: Use of Combat Simulators

The Secretary of the Army shall expand the use of simulations, simulators, and advanced training devices and technologies in order to increase training opportunities for ARNG members and units.

Subtitle B—Assessment of National Guard Capability

Section 1121: Deployability Rating System

The Secretary of the Army shall modify the readiness rating system for units of the Army Reserve and Army National Guard to ensure that it provides an accurate assessment of the deployability of a unit and any shortfalls that require the provision of additional resources. In making such modifications, the Secretary shall ensure that the unit readiness rating system is designed so

(1) That the personnel readiness rating of a unit reflects
   a. Both the percentage of the unit's authorized strength that is manned and deployable and the fill and deployability rate for critical occupational specialties necessary for the unit to carry out its basic mission requirements; and
   b. The number of unit personnel who are qualified in their primary MOS; and

(2) That the equipment readiness assessment of a unit
   a. Documents all equipment required for deployment;
   b. Reflects only the equipment that is directly possessed by the unit;
   c. Specifies the effect of substitute items; and
   d. Assesses the effect of missing components and sets on the readiness of major equipment items.

Section 1122: Inspections

Amends Title 32, section 105, of U.S. Code as follows:

(a) Under regulations prescribed by him, the Secretary of the Army shall have an inspection made by inspectors general, or, if necessary, by any other commissioned officers of the Regular Army detailed for that purpose, to determine whether—

(1) The amount and condition of property held by the Army National Guard are satisfactory;

(2) The Army National Guard is organized as provided in this title;

(3) The members of the Army National Guard meet prescribed physical and other qualifications;

(4) The Army National Guard and its organization are properly uniformed, armed, and equipped and are being trained and instructed for active duty in the field, or for coast defense;

(5) Army National Guard records are being kept in accordance with this title;

(6) The accounts and records of each property and fiscal officer are being properly maintained; and

(7) The units of the Army National Guard meet requirements for deployment.

The Secretary of the Air Force has a similar duty with respect to the Air National Guard.

(b) The reports of inspections under subsection (a) are the basis for determining whether the National Guard is entitled to the issue of military property as authorized under this title and to retain that property; and for determining which organizations and persons constitute units and members of the National Guard; and for determining which units of the National Guard meet deployability standards.

Subtitle C—Compatibility of Guard Units with Active Component Units

Section 1131: Active Duty Associate Unit Responsibility

(a) The Secretary of the Army shall require that each ARNG combat unit be associated with an active-duty combat unit.

(b) The commander (at brigade level or higher) of the associated active duty unit shall be responsible for:

(1) Approving the training program of the National Guard unit;

(2) Reviewing the readiness report of the National Guard unit;

(3) Assessing the manpower, equipment, and training resource requirements of the National Guard unit; and

(4) Validating, not less often than annually, the compatibility of the National Guard unit with the active duty forces.

(c) The Secretary shall begin to implement subsection (a) during fiscal year 1993 and shall achieve full implementation of the plan by October 1, 1995.

Section 1132: Training Compatibility

Amends section 414 of the NDAA for 1992 and 1993 by adding the following new paragraph:

After September 30, 1994, not less than 3,000 warrant officers and enlisted members [in addition to the 2,000 officers] shall be assigned to serve as advisers under the program.

Section 1133: Systems Compatibility

(a) The Secretary of the Army shall develop and implement a program to ensure that Army personnel systems, supply systems, maintenance management systems, and finance systems are compatible across all Army components.

(b) Not later than September 30, 1993, the Secretary shall submit to the House and Senate Armed Services Committees a report describing this program and setting forth a plan to implement the program by the end of fiscal year 1997.

Section 1134: Equipment Compatibility

Amends Title 10, section 10541(b), of U.S. Code, which describes the National Guard and Army Reserve Equipment annual report to Congress, by adding the following paragraph:

(8) A statement of the current status of the compatibility of equipment between the Army reserve components and active forces, the effect of that level of incompatibility on combat effectiveness, and a plan to achieve full equipment compatibility.

Section 1135: Deployment Planning Reform

(a) The Secretary of the Army shall develop a system for identifying the priority for mobilization of RC units. The priority system shall be based on regional contingency planning requirements and doctrine to be integrated into the Army war planning process.

(b) The system shall include the use of Unit Deployment Designators to specify the post-mobilization training days allocated to a unit before deployment. The Secretary shall specify standard designator categories in order to group units according to the timing of deployment after mobilization.

(c) Use of Designators

  (1) The Secretary shall establish procedures to link the Unit Deployment Designator system to the process by which resources are provided for ARNG units.

  (2) The Secretary shall develop a plan that allocates greater funding for training, full-time support, equipment, and manpower in excess of 100 percent of authorized strength to units that have fewer post-mobilization training days.

  (3) The Secretary shall establish procedures to identify the command level at which combat units would, upon deployment, be integrated with AC forces consistent with the Unit Deployment Designator system.

Section 1136: Qualification for Prior-Service Enlistment Bonus

Amends Title 37, section 308i, of U.S. Code to specify that a prior-service enlistment bonus can only be paid if "the specialty associated with the position the member is projected to occupy is a specialty in which the member successfully served while on active duty and attained a level of qualification commensurate with the member's grade and years of service."

Section 1137: Study of Implementation for All Reserve Components

The Secretary of Defense shall conduct an assessment of the feasibility of implementing the provisions of this title for all reserve components. Not later than December 31, 1993, the Secretary shall submit to the House and Senate Armed Services Committees a report containing a plan for such implementation.

National Defense Authorization Act for 1994 (PL 103-160)

Section 515: Active Component Support for Reserve Training

(a) The Secretary of the Army shall, not later than September 30, 1995, establish one or more AC units with the primary mission of providing training support to reserve units. Each such unit shall be part of the active Army force structure and shall have a commander who is on the active-duty list of the Army.

(b) During fiscal year 1994, the Secretary shall submit to the House and Senate Armed Services Committees a plan to meet the requirement in subsection (a). The plan shall include a proposal for any statutory changes that the Secretary considers to be necessary for the implementation of the plan.

Section 517: Revisions to Pilot Program for Active Component Support of the Reserves

(a) Amends section 414, subsection (c), of the NDAA for 1992 and 1993 to read as follows: "The Secretary shall assign not less than 2,000 active component personnel to serve as advisers under the program. After September 30, 1994, the number under the preceding sentence shall be increased to not less than 5,000."

(b) Annual Report on Implementation

    (1) The Secretary of the Army shall include in the annual report to Congress known as the Army Posture Statement a presentation relating to the implementation of the Pilot Program for Active Component Support of the Reserves.

    (2) Each such presentation shall include, with respect to the period covered by the report, the following information:

        a. The promotion rate for officers within the promotion zone who are serving as AC advisers to RC units, compared with the promotion rate for other officers within the promotion zone in the same pay grade and the same competitive category.

        b. The promotion rate for officers below the promotion zone who are serving as AC advisers to RC units, compared in the same manner as specified in subparagraph a.

Section 521: Annual Report on Implementation of Army National Guard Combat Readiness Reform Act

Amends Title 10, section 10542, of U.S. Code by adding

(a) The Secretary of the Army shall include in the annual report to Congress known as the Army Posture Statement a detailed presentation concerning the Army National Guard, including particularly information relating to the implementation of Title XI of PL 102-484, the Army National Guard Combat Readiness Reform Act of 1992 (ANGCRRA).

(b) Each presentation under subsection (a) shall include, with respect to the period covered by the report, the following information concerning the Army National Guard:

   (1) The number and percentage of officers with at least two years of active duty before becoming a member of the Army National Guard.

   (2) The number and percentage of enlisted personnel with at least two years of active duty before becoming a member of the Army National Guard.

   (3) The number of officers who are graduates of one of the service academies and were released from active duty before the completion of their active-duty service obligation and, of those officers—

      a. The number who are serving the remaining period of their active-duty service obligation as a member of the Selected Reserve pursuant to section 1112(a)(1) of ANGCRRA; and

      b. The number for whom waivers were granted by the Secretary under section 1112(a)(2) of ANGCRRA, together with the reason for each waiver.

   (4) The number of officers who were commissioned as distinguished ROTC graduates and were released from active duty before the completion of their active-duty service obligation and, of those officers—

      a. The number who are serving the remaining period of their active-duty service obligation as a member of the Selected Reserve pursuant to section 1112(a)(1) of ANGCRRA; and

b.   The number for whom waivers were granted by the Secretary under section 1112(a)(2) of ANGCRRA, together with the reason for each waiver.

(5)   The number of officers who are graduates of the ROTC program and who are performing their minimum period of obligated service in accordance with section 1112(b) of ANGCRRA by a combination of (A) two years of active duty, and (B) the remainder of their service obligation in the National Guard and, of those officers, the number for whom permission was granted during the preceding fiscal year.

(6)   The number of officers for whom recommendations were made during the preceding fiscal year for a unit vacancy promotion to a grade above first lieutenant and, of those recommendations, the number and percentage that were concurred by an active duty officer under section 1113(a) of ANGCRRA, shown separately for each of the three categories of officers set forth in section 1113(b) of ANGCRRA.

(7)   The number of waivers during the preceding fiscal year under section 1114(a) of ANGCRRA of military education requirements for NCOs and the reason for each such waiver.

(8)   The number and distribution by grade, shown for each state, of personnel in the initial entry training and nondeployability personnel accounting category established under section 1115 of ANGCRRA.

(9)   The number of ARNG members, shown for each state, that were discharged during the previous fiscal year pursuant to section 1115(c)(1) of ANGCRRA for not completing the minimum training required for deployment within 24 months after entering the National Guard.

(10) The number of waivers, shown for each state, that were granted by the Secretary during the previous fiscal year

under section 1115(c)(2) of ANGCRRA, together with the reason for each waiver.

(11) The number of members, shown for each state, who were screened during the preceding fiscal year to determine whether they meet minimum physical profile standards required for deployment and, of those members—

    a. The number and percentage who did not meet minimum physical profile standards required for deployment; and

    b. The number and percentage who were transferred pursuant to section 1116 of ANGCRRA to the personnel accounting category described in paragraph (8).

(12) The number of ARNG members, and the percentage of total membership, shown for each state, who underwent a medical screening during the previous fiscal year as provided in section 1117 of ANGCRRA.

(13) The number of ARNG members, and the percentage of total membership, shown for each state, who underwent a dental screening during the previous fiscal year as provided in section 1117 of ANGCRRA.

(14) The number of ARNG members, and the percentage of total membership, shown for each state, over the age of 40 who underwent a full physical examination during the previous fiscal year for purposes of section 1117 of ANGCRRA.

(15) The number of ARNG units that are scheduled for early deployment in the event of a mobilization, and, of those units, the number that are dentally ready for deployment in accordance with section 1118 of ANGCRRA.

(16) The estimated post-mobilization training time for each ARNG combat unit, and a description, displayed in broad categories and by state, of what training would need to be accomplished for ARNG combat units in a postmobilization period for purposes of section 1119 of ANGCRRA.

(17) A description of the measures taken during the preceding fiscal year to comply with the requirement in section 1120 of ANGCRRA to expand the use of simulations, simula-

tors, and advanced training devices and technologies for ARNG members and units.

(18) Summary tables of unit readiness, shown for each state, and drawn from the unit readiness rating system as required by section 1121 of ANGCRRA, including the personnel readiness rating information and the equipment readiness assessment information required by that section, together with—

    a. Explanations of the information shown in the table; and

    b. The Secretary's overall assessment of the deployability of ARNG units, including a discussion of personnel deficiencies and equipment shortfalls in accordance with section 1121.

(19) Summary tables, shown for each state, of the results of inspections of ARNG units by inspectors general or other commissioned officers of the Regular Army under the provisions of section 105 of title 32, together with explanations of the information shown in the tables, and including display of—

    a. The number of such inspections;

    b. Identification of the entity conducting each inspection;

    c. The number of units inspected; and

    d. The overall results of such inspections, including the inspector's determination of whether the unit met deployability standards, and for those units not meeting deployability standards, the reasons for such failure and the status of corrective actions.

(20) A listing, for each ARNG combat unit, of the active-duty combat unit associated with it in accordance with section 1131(a) of ANGCRRA, shown by state and accompanied by—

    a. The AC commander's assessment of the manpower, equipment, and training resource requirements of the ARNG unit in accordance with section 1131(b)(3) of ANGCRRA; and

    b. The results of the AC commander's validation of the compatibility of the ARNG unit with active duty forces in accordance with section 1131(b)(4) of ANGCRRA.

(21) A specification of the active-duty personnel assigned to RC units pursuant to section 414(c) of the NDAA for 1992 and 1993, shown by state, by rank of officers, warrant officers, and enlisted members, and by unit or other organizational entity of assignment.

(c) The requirement to include information under any paragraph in subsection (b) shall take effect in the year following the year in which that provision of ANGCRRA has taken effect. Before then, the Secretary shall include any information that may be available covering that topic.

(d) In this section, the term "state" includes the District of Columbia, Puerto Rico, Guam, and the Virgin Islands.

National Defense Authorization Act for 1995 (PL 103-337)

Section 413 amends section 414(c) of the NDAA for 1992 and 1993 to delay the increase in AC personnel assigned to RC units from September 30, 1994, to September 30, 1996.

Section 516 amends section 1111 of ANGCRRA (Title XI of PL 102-484) by adding the following new subsection:

On a semiannual basis, the Secretary of the Army shall furnish to the Chief of the National Guard Bureau a list containing the name, home of record, and last-known mailing address of each Army officer who during the previous six months was honorably separated from active duty in the grade of major or below.

Section 521: Sense of Congress Concerning the Training and Modernization of the Reserve Components

(a) Congress makes the following findings:

(1) The force structure specified in the report resulting from the Bottom-Up Review conducted by DoD during 1993

assumes increased reliance on the reserve components of the Armed Forces.

(2) The mobilization of the reserve components for the Persian Gulf War was handicapped by shortfalls in training, readiness, and equipment.

(3) The mobilization of the Army reserve components for the Persian Gulf War was handicapped by a lack of a standard readiness evaluation system, which resulted in a lengthy reevaluation of training and equipment readiness of Army National Guard and Army Reserve units before they could be deployed.

(4) Funding and scheduling constraints continue to limit the opportunity for ARNG combat units to carry out adequate maneuver training.

(5) Funding constraints continue to handicap the readiness and modernization of the reserve components and their interoperability with the active forces.

(b) It is the sense of Congress that the Secretary of Defense, with the advice and assistance of the Chairman of the Joint Chiefs of Staff, should establish—

(1) A standard readiness evaluation system that is uniform for all forces within each military service; and

(2) A standard readiness rating system that is uniform for the military department.

(c) It is the sense of Congress that the Secretary of Defense should assess the budget submission of each military department each year to determine (taking into consideration the advice of the Chairman of the Joint Chiefs of Staff) the extent to which National Guard and reserve units would, under that budget submission, be trained and modernized to the standards needed for them to carry out the full range of missions required of them under current DoD plans. Based upon such assessment each year, the Secretary should adjust the budget submissions of the military departments as necessary in order to meet the priorities established by the Secretary of Defense for the total force.

National Defense Authorization Act for 1996 (PL 104-106)

Section 413: Counting of Certain Active Component Personnel Assigned in Support of Reserve Component Training
Amends section 414(c) of the NDAA for 1992 and 1993 by adding the following new paragraph:

(2) The Secretary of Defense may count toward the number of AC personnel required to be assigned to serve as advisers any AC personnel who are assigned to an AC unit (A) that was established principally for the purpose of providing dedicated training support to RC units; and (B) the primary mission of which is to provide such dedicated training support.

Section 514: Revisions to Army Guard Combat Reform Initiative to Include Army Reserve Under Certain Provisions and Make Certain Revisions

(a) Amends section 1111 of the Army National Guard Combat Readiness Reform Act of 1992 (Title XI of PL 102-484) by striking out subsections (a) and (b) setting goals for the percentage of officers and enlisted personnel with AC experience, and replacing them with the following:

(1) The Secretary of the Army shall increase the number of qualified prior active-duty officers in the ARNG by providing a program that permits the separation of officers on active duty with at least two, but less than three, years of active service upon condition that the officer is accepted for appointment in the Army National Guard. The Secretary shall have a goal of having not fewer than 150 officers become ARNG members each year under this section.

(2) The Secretary of the Army shall increase the number of qualified prior active-duty enlisted members in the ARNG through the use of enlistments described in section 8020 of the Department of Defense Appropriations Act, 1994

(PL 103-139).[2] The Secretary shall enlist not fewer than 1,000 new enlisted members each year under enlistments described in that section.

(b) Amends section 1112(b) of ANGCRRA (Service in the Selected Reserve in Lieu of Active Duty Service for ROTC Graduates) by striking out "National Guard" and inserting "Selected Reserve."

(c) Amends section 1113 of ANGCRRA (Review of Officer Promotions) by
  (1) In subsection (a), striking out "National Guard" and inserting "Selected Reserve."
  (2) Replacing subsection (b) with: Subsection (a) applies to officers in all units of the Selected Reserve that are designated as combat units or that are designated for deployment within 75 days of mobilization. Subsection (a) shall take effect 90 days after the enactment of the NDAA for 1996.

(d) Amends section 1115 of ANGCRRA (Initial Entry Training and Nondeployable Personnel) by striking out "National Guard" and inserting "Selected Reserve."

(e) Amends section 1116 of ANGCRRA (Accounting of Members Who Fail Physical Deployability Standards) by striking out "National Guard" and inserting "Selected Reserve."

(f) Amends section 1120 of ANGCRRA (Use of Combat Simulators) by adding "and the Army Reserve" before the period at the end.

Section 515: Active Duty Associate Unit Responsibility

(a) Amends section 1131(a) of the Army National Guard Combat Readiness Reform Act of 1992 (Title XI of PL 102-484) as follows: The Secretary of the Army shall require—

---

[2]   Section 8020 refers to

members in combat arms skills or to members who enlist in the armed services on or after July 1, 1989, under a program continued or established by the Secretary of Defense in fiscal year 1991 to test the cost-effective use of special recruiting incentives involving not more than nineteen noncombat arms skills approved in advance by the Secretary of Defense.

(1) That each ARNG ground combat maneuver brigade that (as determined by the Secretary) is essential for the execution of the National Military Strategy be associated with an active-duty combat unit; and

(2) That Army Selected Reserve CS and CSS units that (as determined by the Secretary) are essential for the execution of the National Military Strategy be associated with active-duty units.

(b) Section 1131(b) is amended by striking out "National Guard combat unit" and inserting "National Guard unit or Army Selected Reserve unit that (as determined by the Secretary) is essential for the execution of National Military Strategy."

Section 704: Medical and Dental Care for Members of the Selected Reserve Assigned to Early Deploying Units of the Army Selected Reserve

(a) Amends Title 10, section 1074d, by adding the following new subsection:

(1) The Secretary of the Army shall provide to members of the Army Selected Reserve who are assigned to units scheduled for deployment within 75 days after mobilization the following medical and dental services:

a. An annual medical screening.

b. For members who are over 40 years of age, a full physical examination not less often than once every two years.

c. An annual dental screening.

d. The dental care identified in an annual dental screening as required to ensure that a member meets the dental standards required for deployment in the event of a mobilization.

(2) The services provided under this subsection shall be provided at no cost to the member.

(b) Repeals sections 1117 and 1118 of the Army National Guard Combat Readiness Reform Act of 1992 (Title XI of PL 102-484).

National Defense Authorization Act for 1997 (PL 104-201)

Section 545: Report on Number of Advisers in Active Component
Support of Reserves Pilot Program

  (a) Not later than six months after the enactment of this Act, the
Secretary of Defense shall submit to the House and Senate
National Security Committees a report setting forth the Secretary's determination as to the appropriate number of AC personnel to be assigned as advisers to reserve components under
section 414 of the NDAA for 1992 and 1993. If the Secretary's
determination is that such number should be different from the
required minimum number in subsection (c), the Secretary shall
provide a justification for the number recommended.

  (b) Amends 10 USC 12001 by striking out, "During fiscal years
1992 and 1993, the Secretary of the Army shall institute," and
inserting "The Secretary of the Army shall carry out."

National Defense Authorization Act for 2000 (PL 106-65)

Section 1066(d)(2) amends section 414 of the NDAA for 1992 and
1993 and 10 USC 12001 by removing the term "pilot" and increasing
the number of AC advisers from 2,000 to 5,000.

National Defense Authorization Act for 2005 (PL 108-375)

Section 513 establishes the Commission on the National Guard and
Reserves and describes its composition and duties.

Section 515: Army Program for Assignment of AC Advisers to Units of
the Selected Reserve

  (a) Amends section 414(c)(1) of the NDAA for 1992 and 1993 and
10 USC 12001 striking out "5,000" and inserting "3,500."

  (b) Notwithstanding the amendment made by subsection (a), the
Secretary of the Army may not reduce the number of AC reserve

support personnel below the existing number until the report required by subsection (c) has been submitted.

(c) Not later than March 31, 2005, the Secretary of the Army shall submit to the House and Senate Armed Services Committees a report on the support by active components of the Army for training and readiness of the Army National Guard and Army Reserve. The report shall include an evaluation and determination of each of the following:

(1) The effect on the ability of the Army to improve such training and readiness resulting from the reduction in subsection (a) in the minimum number of AC reserve support personnel.

(2) The adequacy of having 3,500 members of the Army assigned as AC reserve support personnel in order to meet emerging training requirements in the Army reserve components in connection with unit and force structure conversions and preparations for wartime deployment.

(3) The nature and effectiveness of efforts by the Army to reallocate the 3,500 personnel assigned as AC reserve support personnel to higher priority requirements and to expand the use of reservists on active duty to meet RC training needs.

(4) Whether the Army is planning further reductions in the number of AC reserve support personnel and, if so, the scope and rationale for those reductions.

(5) Whether an increase in Army RC full-time support personnel will be required to replace the loss of AC reserve support personnel.

(d) In this section, the term "AC reserve support personnel" means the AC Army personnel assigned as advisers to RC units pursuant to section 414 of the NDAA for 1992 and 1993.

Section 1043: Report on Training Provided to Members of the Armed Forces to Prepare for Post-Conflict Operations

(a) The Secretary of Defense shall conduct a study to determine the extent to which members of the Armed Forces assigned to duty in support of contingency operations receive training in preparation for post-conflict operations and to evaluate the quality of such training.

(b) As part of the study under subsection (a), the Secretary shall specifically evaluate the following:

 (1) The doctrine, training, and leader-development system necessary to enable members of the Armed Force to successfully operate in post-conflict operations.

 (2) The adequacy of the curricula at military educational facilities to ensure that the Armed Forces has a cadre of members skilled in post-conflict duties, including a familiarity with applicable foreign languages and foreign cultures.

 (3) The training time and resources available to members and units of the Armed Forces to develop awareness about ethnic backgrounds, religious beliefs, and political structures of the people living in areas in which the Armed Forces operate and areas in which post-conflict operations are likely to occur.

 (4) The adequacy of training transformation to emphasize post-conflict operations, including interagency coordination in support of combatant commanders.

(c) Not later than May 1, 2005, the Secretary shall submit to the House and Senate Armed Forces Committees a report on the result of the study conducted under this section.

U.S. Code

Relatively few of the provisions related to Title XI are in U.S. Code. However, three provisions are included as notes to 10 USC 12001 (Authorized Strengths: Reserve Components). The first is based on section 414 of the NDAA for 1992 and 1993, as amended.

Program for Active Component Support of the Reserves

(a) The Secretary of the Army shall carry out a program to provide AC advisers to combat units, CS units, and CSS units in the

Selected Reserve that have a high priority for deployment on a time-phased troop deployment list or have another contingent high priority for deployment. The advisers shall be assigned to full-time duty in connection with organizing, administering, recruiting, instructing, or training such units.

(b) The objectives of the program are as follows:

(1) To improve the readiness of units in the Army reserve components.

(2) To increase substantially the number of AC personnel directly advising RC unit personnel.

(3) To provide a basis for determining the most effective mix of RC and AC personnel in organizing, administering, recruiting, instructing, or training RC units

(4) To provide a basis for determining the most effective mix of AC officers and enlisted personnel in advising RC units regarding organizing, administering, recruiting, instructing, or training RC units.

(c) Personnel to be assigned:

(1) The Secretary shall assign not less than 3,500 AC personnel to serve as advisers under the program.

(2) The Secretary of Defense may count toward the number of AC personnel required to be assigned to serve as advisers any AC personnel who are assigned to an AC unit (A) that was established principally for the purpose of providing dedicated training support to RC units; and (B) the primary mission of which is to provide such dedicated training support

(d) Based on the experience under the pilot program, the Secretary of the Army shall by April 1, 1993, submit to the House and Senate Armed Services Committees a report containing the Secretary's evaluation of the program to date. As part of the budget submission for fiscal year 1995, the Secretary shall submit any recommendations for expansion or modification of the program, together with a proposal for any statutory changes that the Secretary considers necessary to implement the program on a permanent basis. In no case may the number of AC personnel assigned to the program decrease below the number specified for the pilot program.

The second note, "Annual report on implementation of the Pilot Program for Active Component Support of the Reserves," includes the language from section 517(b) of the NDAA for 1994 requiring the Army Posture Statement to include information on the promotion rates of AC officers assigned as advisers to RC units in comparison to other AC officers. The third note, "Assignment of active component advisers to units of Selected Reserve; limitation on reductions; report; definition," includes the language from section 515(b)–(d) requiring the Secretary of the Army to submit a report on the reduction of AC advisers from 5,000 to 3,500 by March 31, 2005.

Language from Title XI, section 1122, on inspections of National Guard units can be found in 32 USC 105. Section 1136 of Title XI, which sets conditions on prior-service enlistment bonuses, can be found in Title 37, section 308i(a). The National Guard and reserve component equipment report to Congress, which was modified by Title XI, section 1134, is described in 10 USC 10541. The information added to the Army Posture Statement by section 521(a)–(b) of the NDAA for 1994 can be found in 10 USC 10542. Changes made to medical and dental screening requirements by section 704 of the NDAA for 1996, which also repealed sections 1117 and 1119 of Title XI, are in 10 USC 1074a.

Abbreviations

AAR	after-action review
AC	active component
AGR	Active Guard and Reserve
ANGCRRA	Army National Guard Combat Readiness Reform Act of 1992
AR	Army regulation
ARFORGEN	Army Force Generation
ARNG	Army National Guard
ASA(M&RA)	Assistant Secretary of the Army for Manpower and Reserve Affairs
AT	annual training
BCT	brigade combat team
CALL	Center for Army Lessons Learned
Co-ADOS	Contingency Active Duty for Operational Support
CONUSA	Continental U.S. Army
CS	combat support
CSS	combat service support
DAIG	Department of the Army Inspector General
DMDC	Defense Manpower Data Center

DoD	Department of Defense
EXORD	execution order
FOB	forward operating base
FORSCOM	U.S. Army Forces Command
FTS	full-time support
FY	fiscal year
GAO	General Accounting Office (now Government Accountability Office)
HQDA	Headquarters, Department of the Army
IED	improvised explosive device
IPR	in-process review
LSBn	logistics support battalion
METL	mission-essential task list
MOS	military occupational specialty
MRAP	mine-resistant, ambush-protected vehicle
MSAD	mobilization station arrival date
NCO	noncommissioned officer
NDAA	National Defense Authorization Act
NGB	National Guard Bureau
ODS	Operation Desert Shield/Operation Desert Storm
OEF	Operation Enduring Freedom
OIF	Operation Iraqi Freedom
OND	Operation New Dawn
OPLAN	operation plan
PL	Public Law
PTAE	premobilization training assistance element

RC	reserve component
ROTC	Reserve Officer Training Corps
RTC	regional training center
SRP	Soldier Readiness Processing
TADSS	training aids, devices, simulators, and simulations
TRADOC	U.S. Army Training and Doctrine Command
TSB	training support brigade
TSBn	training support battalion
TSD	training support division
TTHS	trainees, transients, holdees, and students
USAR	U.S. Army Reserve
USARC	U.S. Army Reserve Command
USC	U.S. Code
USR	unit status report

Bibliography

1398th Deployment Support Brigade, "After Action Review," briefing, September 13, 2013.

155th Brigade Combat Team, "After Action Review," briefing, October 13, 2009.

157th Infantry Brigade, "After Action Review," briefing, June 16, 2010.

1-6 Kansas Agribusiness Development Team, "After Action Review," briefing, July 12, 2009.

AR—*See* Army Regulation.

Army Regulation 11-30, "Army WARTRACE Program," Washington, D.C.: Department of the Army, July 28, 1995.

Army Regulation 220-1, "Army Unit Status Reporting and Force Registration—Consolidated Policies," Washington, D.C.: Department of the Army, August 30, 1988.

Army Regulation 220-1, "Army Unit Status Reporting and Force Registration—Consolidated Policies," Washington, D.C.: Department of the Army, April 15, 2010.

Army Regulation 350-1, "Army Training and Leader Development," Washington, D.C.: Department of the Army, August 4, 2011.

Army Regulation 525-29, "Army Force Generation," Washington, D.C.: Department of the Army, March 14, 2011.

Arnold, Richard E., "Active Component Support to Reserve Component Training, Changes to Training Support XXI," Carlisle Barracks, Pa.: U.S. Army War College, April 7, 2003.

Association of the United States Army, "First Army: Training for Today's Requirements and Tomorrow's Contingencies," April 2012. As of March 11, 2014: http://www.ausa.org/publications/torchbearercampaign/torchbearerissuepapers/Documents/TBIP_first_army_web.pdf

Brauner, Marygail K., and Glenn A. Gotz, *Manning Full-Time Positions in Support of the Selected Reserve*, Santa Monica, Calif.: RAND Corporation, R-4034-RA, 1991. As of September 18, 2014:
http://www.rand.org/pubs/reports/R4034.html

Brauner, Marygail, Harry Thie, and Roger Brown, *Assessing the Structure and Mix of Future Active and Reserve Forces: Effectiveness of Total Force Policy During the Persian Gulf Conflict*, Santa Monica, Calif.: RAND Corporation, MR-132-OSD, 1992. As of September 18, 2014:
http://www.rand.org/pubs/monograph_reports/MR132.html

Broomall, Hugh T., "Integration of the National Guard into the Total Force," Ft. Leavenworth, Kan.: School of Advanced Military Studies, U.S. Army Command and General Staff College, 1992.

Buchalter, Alice R., and Seth Elan, *Historical Attempts to Reorganize the Reserve Components*, Washington, D.C.: U.S. Library of Congress, Federal Research Division, October 2007.

Burns, Brad K. "Operations Brigade S3 Replaced by Operations Battalion," Ft. Leavenworth, Kan.: School of Advanced Military Studies, U.S. Army Command and General Staff College, October 25, 2012.

CALL—*See* Center for Army Lessons Learned.

Camp Atterbury, "IRDO [Individual Replacement Deployment Operations] Training Tasks," briefing, July 7, 2014a. As of October 1, 2014:
http://www.atterburymuscatatuck.in.ng.mil/Portals/18/PageContents/Training/IRDO/IRDO%20Required%20Training%20Tasks_07JUL2014_Web.pdf

———, "Online Training Pre-Requisites Checklist," July 21, 2014b. As of October 1, 2014:
http://www.atterburymuscatatuck.in.ng.mil/Portals/18/PageContents/Training/IRDO/Annex%20IV%20-%20Online%20Training%20Prerequisites%20Checklist_22JULY2014.pdf

Center for Army Lessons Learned, *The First 100 Days: Operation Enduring Freedom*, Ft. Leavenworth, Kan.: Combined Arms Center, November 2008.

———, *39th Infantry Brigade Combat Team: Mobilization and Deployment Journal*, Ft. Leavenworth, Kan.: Combined Arms Center, June 2009.

Chapman, Dennis, "Planning for Employment of the Reserve Components: Army Practice, Past and Present," Arlington, Va.: Institute of Land Warfare, Association of the United States Army, National Security Affairs Paper No. 69, September 2008.

Combined Joint Task Force Phoenix, "After Action Review," briefing, October 19, 2009.

Commission on the National Guard and Reserves, *Commission on the National Guard and Reserves: Transforming the National Guard and Reserves into a 21st Century Operational Force*, final report to Congress and the Secretary of Defense, January 31, 2008.

DAIG—*See* Department of the Army Inspector General.

Defense Science Board Task Force, *Deployment of Members of the National Guard and Reserve in the Global War on Terrorism*, Washington, D.C.: Office of the Under Secretary of Defense for Acquisition, Technology, and Logistics, 2007.

Department of the Army Inspector General, *Special Assessment of Army National Guard Brigades' Mobilization*, Washington, D.C., June 1991.

Doubler, Michael D., and Vance Renfroe, "The National Guard and the Total Force Policy," *The Modern National Guard*, Tampa, Fla.: Faircount LLC, 2003, pp. 42–47.

First Army, "History," undated. As of February 25, 2014:
http://www.first.army.mil/content.aspx?ContentID=200

———, "Training for Today's Requirements and Tomorrow's Contingencies: 2012–2014," Rock Island, Ill., 2013a. As of February 25, 2014:
http://www.first.army.mil/documents/pdf/1A_Brochure_2013_Version.pdf

———, "First Army," briefing, February 13, 2013b.

———, "First Army Response to the Draft Rand Report, Active Component Responsibility in Reserve Component Pre- and Post-Mobilization Training, March 2014," May 12, 2014.

Flores, Pascual, "RTC-East 'Ends as It Began,'" Joint Base McGuire-Dix-Lakehurst, N.J.: Public Affairs, August 30, 2012. As of February 27, 2014:
http://www.jointbasemdl.af.mil/news/story.asp?id=123316258

FORSCOM—*See* U.S. Army Forces Command.

GAO—*See* U.S. General Accounting Office.[1]

Gates, Robert M., Secretary of Defense, *Utilization of the Total Force*, memorandum, *Department of Defense*, Washington, D.C., January 19, 2007.

Goldich, Robert L., *The Army's Roundout Concept After the Persian Gulf War*, Washington, D.C.: Congressional Research Service, 91-763 F, October 22, 1991.

Headquarters, Department of the Army, "Soldier's Manual of Common Tasks: Warrior Skills Level 1," STP 21-1-SMCT, April 2014.

---

[1]   Now known as the U.S. Government Accountability Office.

Joint Publication 1-02, "Dictionary of Military and Associated Terms," Washington, D.C.: U.S. Department of Defense, November 8, 2010 (as amended through March 15, 2013). As of April 12, 2013:
http://www.dtic.mil/doctrine/new_pubs/jp1_02.pdf

Lippiatt, Thomas F., and J. Michael Polich, *Reserve Component Unit Stability: Effects on Deployability and Training*, Santa Monica, Calif.: RAND Corporation, MG-954-OSD, 2010. As of September 18, 2014:
http://www.rand.org/pubs/monographs/MG954.html

———, *Leadership Stability in Army Reserve Component Units*, Santa Monica, Calif.: RAND Corporation, MG-1251-OSD, 2013. As of September 18, 2014:
http://www.rand.org/pubs/monographs/MG1251.html

MacCarley, Mark, "The Reserve Component: Trained and Ready? Lessons of History," *Military Review*, Vol. 92, No. 3, May–June 2012, pp. 35–46.

McHugh, John M., Secretary of the Army, "Army Directive 2012-08 (Army Total Force Policy)," September 4, 2012.

———, Secretary of the Army, "Total Army Training Validation Integrated Planning Team Charter," March 12, 2013.

McHugh, John M., and Raymond T. Odierno, *2012 Army Posture: The Nation's Force of Decisive Action*, Addendum F—Reserve Component Readiness, submitted to the U.S. Senate and House of Representatives, February 17, 2012.

Miller, Thomas G., Deputy Chief of Staff, G-3/5/7, U.S. Army Forces Command, "Title XI Reduction Implementation Guidance," July 29, 2005.

National Guard Bureau, "Implementing the Army Force Generation Model in the Army National Guard: A Formula for Operational Capacity," white paper, August 1, 2011. As of March 11, 2014:
http://www.arng.army.mil/News/publications/Publications/ARFORGENwhitePaper1aug2011v3g2g.pdf

Pernin, Christopher G., Dwayne M. Butler, Louay Constant, Lily Geyer, Duncan Long, Dan Madden, John E. Peters, James D. Powers, and Michael Shurkin, *Readiness Reporting for an Adaptive Army*, Santa Monica, Calif.: RAND Corporation, RR-230-A, 2013. As of September 18, 2014:
http://www.rand.org/pubs/research_reports/RR230.html

PL—*See* Public Law.

Public Law 102-190, National Defense Authorization Act for Fiscal Years 1992 and 1993, Section 414, Pilot Program for Active Component Support of the Reserves, December 5, 1991.

Public Law 102-484, National Defense Authorization Act for Fiscal Year 1993, Title XI, Army National Guard Combat Readiness Reform Act of 1992, October 23, 1992.

Public Law 103-139, Department of Defense Appropriations Act, 1994, Section 8020, November 11, 1993.

Public Law 103-160, National Defense Authorization Act for Fiscal Year 1994, Section 515, Active Component Support for Reserve Training; Section 517, Revisions to Pilot Program for Active Component Support of the Reserves; and Section 521, Annual Report on Implementation of Army National Guard Combat Readiness Reform Act, November 30, 1993.

Public Law 103-337, National Defense Authorization Act for Fiscal Year 1995, Section 413, Delay in Increase in Number of Active Component Members to Be Assigned for Training Compatibility with Guard Units; Section 516, Semiannual Report on Separations of Active Army Officers; and Section 521, Sense of Congress Concerning the Training and Modernization of the Reserve Components, October 5, 1994.

Public Law 104-106, National Defense Authorization Act for Fiscal Year 1996, Section 413, Counting of Certain Active Component Personnel Assigned in Support of Reserve Component Training; Section 514, Revisions to Army Guard Combat Reform Initiative to Include Army Reserve Under Certain Provisions and Make Certain Revisions; Section 515, Active Duty Associate Unit Responsibility; and Section 704, Medical and Dental Care for Members of the Selected Reserve Assigned to Early Deploying Units of the Army Selected Reserve, February 10, 1996.

Public Law 104-201, National Defense Authorization Act for Fiscal Year 1997, Section 545, Report on Number of Advisers in Active Component Support of Reserves Pilot Program, September 23, 1996.

Public Law 106-65, National Defense Authorization Act for Fiscal Year 2000, Section 1066, Technical and Clerical Amendments, October 5, 1999.

Public Law 108-375, Ronald W. Reagan National Defense Authorization Act for Fiscal Year 2005, Section 513, Commission on the National Guard and Reserves; Section 515, Army Program for Assignment of Active Component Advisers to Units of the Selected Reserve; and Section 1043, Report on Training Provided to Members of the Armed Forces to Prepare for Post-Conflict Operations, October 28, 2004.

Salter, Margaret S., *After Action Reviews: Current Observations and Recommendations*, U.S. Army Research Institute for the Behavioral and Social Sciences, 2007.

Sanzo, Rachel L., "Combat Vets Prepare Guardsmen for War Zone," 42nd Infantry Division Public Affairs, August 15, 2008.

Schuette, Rob, "Regional Training Center-North Standing up at McCoy," *The Real McCoy*, March 14, 2008. As of February 27, 2014: http://www.mccoy.army.mil/ReadingRoom/Newspaper/realmccoy/03142008/Regional_Training_Center_North.htm

Sortor, Ronald E., Thomas F. Lippiatt, J. Michael Polich, and James C. Crowley, *Training Readiness in the Army Reserve Components*, Santa Monica, Calif.: RAND Corporation, MR-474-A, 1994. As of September 24, 2014: http://www.rand.org/pubs/monograph_reports/MR474.html

Thurman, James D., Deputy Chief of Staff, G-3/5/7, "Reserve Component Deployment Expeditionary Force Pre and Post-Mobilization Training Strategy," HQDA EXORD 150-08, February 29, 2008.

U.S. Army, "ARFORGEN 101 17AUG," briefing, August 17, 2012.

U.S. Army Combined Arms Training Center–Training, *Leader's Guide to After Action Reviews (AAR)*, Ft. Leavenworth, Kan., 2011.

U.S. Army Forces Command, "Principles for AC to RC Support," briefing, undated.

———, "Reserve Component Training," FORSCOM/ARNG/USAR Regulation 350-2, Ft. McPherson, Ga., October 27, 1999.

———, "Army Force Generation Model—2013," briefing, March 14, 2012.

U.S. Army Forces Command, G-3/5/7, "Information Paper: Talking Points for Rand Study 'AC Support to RC,'" March 2013.

U.S. Army G-3/5/7, "AR 350-1, Training & Leader Development Information Briefing: Training General Officer Steering Committee, 28–29 November 2012," briefing, November 14, 2012.

U.S. Code, Title 10, Section 12001, Authorized Strengths: Reserve Components, December 1, 1994.

U.S. Code, Title 10, Section 10542, Army National Guard Combat Readiness: Annual Report, February 10, 1996.

U.S. General Accounting Office,[2] *Problems in Implementing the Army's CAPSTONE Programs to Provide All Reserve Components with a Wartime Mission*, Washington, D.C., GAO/FPCD-82-59, September 22, 1982.

———, *National Guard: Peacetime Training Did Not Adequately Prepare Combat Brigades for Gulf War*, Washington, D.C., GAO/NSIAD-91-263, September 1991.

———, *Operation Desert Storm: Army Had Difficulty Providing Adequate Active and Reserve Support Forces*, Washington, D.C., GAO/NSIAD-92-67, March 1992a.

———, *Army Training: Replacement Brigades Were More Proficient Than Guard Round-Out Brigades*, Washington, D.C., GAO/NSIAD-93-4, November 1992b.

———, *Army National Guard: Combat Brigades' Ability to Be Ready for War in 90 Days Is Uncertain*, Washington, D.C., GAO/NSIAD-95-91, June 2, 1995.

---

[2]   Now known as the U.S. Government Accountability Office.

U.S. House of Representatives, *Impact of the Persian Gulf War and the Decline of the Soviet Union on How the United States Does Its Defense Business: Hearings Before the Committee on Armed Services*, Witness Panel 3, March 8, 1991a, pp. 157–236.

———, *Impact of the Persian Gulf War and the Decline of the Soviet Union on How the United States Does Its Defense Business: Hearings Before the Committee on Armed Services*, Witness Panel 16, June 12, 1991b, pp. 915–974.

———, *Hearings on National Defense Authorization Act for Fiscal Year 1993— H.R. 5006 and Oversight of Previously Authorized Programs: Hearings Before the Subcommittee on Military Personnel and Compensation*, Witness Panel 16, May 1, 1992a, pp. 1398–1477.

———, *Hearings on National Defense Authorization Act for Fiscal Year 1993— H.R. 5006 and Oversight of Previously Authorized Programs: Hearings Before the Subcommittee on Military Personnel and Compensation*, Witness Panel 17, May 1, 1992bpp. 1477–1506, .

———, *Regional Threats and Defense Options for the 1990s: Hearings Before the Committee on Armed Services*, Witness Panel 13, May 5, 1992c, pp. 423–473.

———, *National Defense Authorization Act for Fiscal Year 1993: Report of the Committee on Armed Services, House of Representatives on H.R. 5006*, May 19, 1992d.

———, *Reserve and Guard Effectiveness: Hearings Before the Subcommittee on Military Forces and Personnel*, Witness Panel #4, April 21, 1993a, pp. 130–161.

———, *Reserve and Guard Effectiveness: Hearings Before the Subcommittee on Military Forces and Personnel*, Witness Panel #5, April 22, 1993b, pp. 163–195.

U.S. Northern Command, "U.S. Army North," fact sheet, Fort Sam Houston, Tex., May 16, 2013. As of September 18, 2014: http://www.northcom.mil/Newsroom/FactSheets/ArticleView/tabid/3999/ Article/1896/us-army-north.aspx

Weiss, Thomas J., "National Guard Pre-Mobilization Training Certification: 54 Ways to Skin a Cat," Carlisle Barracks, Pa.: U.S. Army War College, March 15, 2008.

White, Carl L., "The Army National Guard and Army Reserve: An Operational Transformation," Carlisle Barracks, Pa.: U.S. Army War College, April 13, 2010.

Witscheber, Loni, "Operation Warrior Trainer Offered at Fort McCoy," October 2, 2006. As of March 7, 2014: http://www.army.mil/article/227/operation-warrior-trainer-offered-at-fort-mccoy